Battle-Wise

Battle-Wise

Seeking Time-Information Superiority in Networked Warfare

David C. Gompert, Irving Lachow,
and Justin Perkins

Foreword by Raymond C. Smith
Afterword by Linton Wells II

Published for the
Center for Technology and National Security Policy
by National Defense University Press
Washington, D.C.
2006

The opinions, conclusions, and recommendations expressed or implied within are those of the individual authors and do not represent the views of other contributors, the Department of Defense or any other agency or department of the Federal Government. This publication is cleared for public release; distribution unlimited.

Portions of this work may be quoted or reprinted without further permission, with credit to the Center for Technology and National Security Policy. A courtesy copy of any reviews and tearsheets would be appreciated.

Library of Congress Cataloging-in-Publication Data

Gompert, David C.
  Battle-wise : seeking time-information superiority in networked warfare / David C. Gompert, Irving Lachow, and Justin Perkins ; foreword by Raymond C. Smith ; afterword by Linton Wells II.
      p. cm.
  Includes bibliographical references and index.
    ISBN 1-57906-072-2 (alk. paper)
  1. Command and control systems—United States. 2. Military art and science—Automation. 3. Computer networks—United States. 4. Operational art (Military science) 5. Information warfare—United States. I. Lachow, Irving. II. Perkins, Justin, 1975– III. National Defense University. Center for Technology and National Security Policy. IV. Title.
UB212.G66 2006
355.3'30410973–dc22

                                                                    2006048202

First Printing, July 2006

NDU CTNSP publications are sold by the U.S. Government Printing Office. For ordering information, call (202) 512-1800 or write to the Superintendant of Documents, U.S. Government Printing Office, Washington, D.C. 20402. For GPO publications on-line access their Web site at: http://bookstore.gpo.gov.

For current publications of the Center for Technology and National Security Policy, consult National Defense University Web site at: http://www.ndu.edu/ctnsp/publications.html.

Contents

Foreword

Rear Admiral Raymond C. Smith, USN (Ret.)

As the world careens into the 21st century, the capacity and means by which the American Armed Forces defend their nation are entering a paradigm-breaking transition period. Previous transitions have been driven by the technologies of weapons and their platforms: from sail, to boiler, to turbine; from foot, to horse, to vehicle; from balloon, to manned aircraft, to unmanned aircraft. This is not so in the 21st century. We need to reset our "warfighting gyro," so to speak. To this end, taking the information revolution as a starting point, *Battle-Wise* argues that only by strengthening the relationship between information technology and brain matter will the U.S. military enhance its ability to outsmart and outfight future adversaries.

The authors approach this transition in great detail by making a strong case for building what they call *battle-wisdom*. An improved light machinegun will not measurably improve our soldiers' capability if they are outsmarted by an adversary who has blended into a hostile town and is not in a uniform. Neither will improved weapons offer the necessary edge against a sophisticated strategic adversary in the information age. What will improve our soldiers' capability are the means to draw discreet tactical information and the capacity to weigh a multitude of tactical options—all at battlefield speed. This is the essence of battle-wisdom.

Building superior battle-wisdom in a competitive world will not be easy. It will require our military to face up to ground-breaking issues that will challenge even the most farsighted of its leaders. For example, platoons and companies will become capable of warfighting decisions that historically have been the responsibility of the battalion. The same could be said for the new battle-wise battalion commander as he relates to his brigade or division. Changes in personnel policy also will have to be contemplated.

Take the concept of lateral entry of potential battle-wise soldiers from analogous civilian occupations. The U.S. Army, Marine Corps, and

Special Operations Forces (SOF) have utilized this avenue for doctors, lawyers, and civil affairs soldiers, but have yet to introduce such changes in recruiting warfighters. Yet metropolitan police officers who spend significant portions of their careers in task forces combating gangs, drugs, or organized crime have relied on the equivalent of battle-wisdom, albeit at less technically sophisticated and violent levels. Their training and experience in domestic urban warfare certainly are more suitable for leadership in the military than are those of the newly certified doctor or lawyer who enters service with only professional education as a credential. The cultural complication confronting this concept is the military rank system. Could a 30-year-old police officer with relevant experience be recruited into service at an advanced rank? Is that police officer qualified to command a platoon of soldiers, especially if they are expected to battle insurgents in an urban environment? What about a company or battalion of soldiers? Such ideas have been floated before; in the context of battle-wisdom, there is a strong case for reexamination.

Traditional service training programs, almost without exception, focus on a conventional adversary. In the 21st century, our military will face not only strategic challengers but also unconventional adversaries who present nondoctrinal and nonstandard threats. How can the military establish and train to a doctrine meant to confront an adversary that has no doctrine or standard operating procedure? The authors' answer is a call for changes in military education and training that are no less sweeping than the technological and geopolitical changes that have fundamentally altered the challenges of warfare.

As commander of the Naval Special Warfare forces during Operations *Desert Shield* and *Desert Storm,* I appreciate how crucial it is for our forces to supplement pre-mission intelligence briefings with real-time ground truth while on mission. Despite their many successes during those campaigns, Navy and other SOF could have been significantly more valuable. To be sure, units and individuals were as well trained as humanly possible, given their 5 months of preparation during Operation *Desert Shield.* Yet in retrospect, what was lacking was the technical capacity to feed updated, collated, and analyzed information by which SOF team leaders could maintain tactical advantage while operating behind enemy lines. And even with the technical means to provide the forces such updates, Navy SOF were not trained to absorb, process, and adapt to a continuous flow of information. Such skills can only be developed through intense training that would have enabled the warfighter to marry the use of instincts with real-time information heretofore unavailable.

Since the first Gulf War—from Somalia to Afghanistan to Iraq—we have learned of the unpredictability and ruthlessness of our present and future adversaries through pain and loss of life. We cannot wait for them to transform into our idea of an adversary because it is not going to happen. Sam Rayburn, former Speaker of the House of Representatives, said it best about learning from mistakes in life: "There is no wisdom in the second kick of a mule." Although he did not have 21st-century warfare in mind, his wisdom is particularly applicable to our military and its challenges.

We certainly have had our challenges in recent military operations. Despite the exceptional heroism, courage, and intellect of today's young military members, we need to reformulate the manner by which we prepare them. For decades we have given them the best weapons systems our country could produce. The authors believe, as I do, that the time has come to augment weapons systems and information networks with the intellectual tools that will enable them to gain and maintain cognitive superiority and thus turn the tables on our clever and nimble adversaries. Only by developing battle-wise soldiers—a daunting, but critically important effort on the part of our military leadership—can we expect to avoid the "second kick of a mule."

Preface

This book is an inquiry into the possibility of improving the operational thinking and decisionmaking of U.S. military individuals, teams, and forces who fight for their nation. The inspiration for this work lies in both the belief that information networking presents a unique chance to improve cognitive effectiveness in battle and the worry that U.S. security interests could suffer if this chance is missed.

The United States is presently unrivalled in military power and assured of remaining so for the foreseeable future, thanks to its resources and the transformation of its forces based on networking principles. However, as adversaries of various sorts and sizes also adopt those same principles and exploit increasingly available and easily usable information technology (IT), U.S. operational advantages and strategic equities could be eroded. The unstoppable spread of information networking and know-how gives rise to the need for a new edge—one that utilizes but transcends networks—by developing people, teams, and decisionmaking methods that convert information into better choices and outcomes. We call this new edge *battle-wisdom*.

We form a view of the need for and nature of battle-wise people, forces, and decisionmaking by tackling several crisscrossing questions:

- As adversaries of the United States acquire IT and employ networking in warfare, how will U.S. operational advantages be affected?
- What options exist to gain new advantages that use but transcend networking?
- How can military decisionmakers make good sense and full use of the flood of information that networks are able to supply?
- How do people think and solve problems in situations of urgency, danger, high stakes, complexity, confusion, and information abundance?
- What are the respective cognitive contributions of reasoning and intuition in such situations, and how are they combined?

■ What are the most valuable cognitive abilities in networked military operations and the new security environment?

■ What policies could improve these key cognitive abilities in military decisionmakers, the better to exploit networked information and cope with complexity?

■ Can such policies give the United States and its democratic allies a new and enduring military advantage?

While the thrust of this volume is toward how people think and decide in battle, it is necessary to examine the conditions under which they work and fight in general and in the new global security environment in particular. This requires exploration of a future of diverse dangers and increasingly networked adversaries. Although this future is approaching with surprising speed, we found scant analysis of the dynamics of hostilities involving two opposed networked forces—a sign that U.S. defense planners are still preoccupied with the need and plans for the networking of U.S. forces. Therefore, we had to postulate such conditions, using China to exemplify a determined, well-resourced, technologically capable potential challenger to U.S. interests and forces, and al Qaeda as a globally distributed, fanatical, stateless terrorist threat.

The selection of these cases is not a casual one. Both China and al Qaeda already realize the leverage that networking offers. China has stepped up the modernization of its force in ways that prefigure a move toward networking, and al Qaeda is busily networking in its own technically simple but operationally cunning ways. U.S. and coalition troops in Iraq are already battling networked terrorists and their insurgent allies, with disquieting results.

Our aim is to understand whether and how advantages in thinking and decisionmaking under operational conditions can affect outcomes—victories or defeats—especially in networked warfare. It is important to identify as precisely as possible the mental abilities, such as anticipation and rapid adaptation, that are of greatest utility in networked operations and thus in strategic competition so that these abilities can be emphasized in the ways that military personnel are recruited, taught, developed, and organized. Still deeper character traits—notably, the willingness to take on responsibility and the propensity to learn—can matter greatly in how and how well soldiers think and decide in the severe and stressful conditions of battle.

We have found it helpful to take an excursion into nonmilitary sectors and organizations, especially some that not only have embraced networking but also recognized that doing so is only a platform for more effective use and performance of people and their minds. The military, a latecomer to the networking revolution, can learn from civilian experience, though it is important to judge carefully what lessons do and do not apply.

Of course, combat presents a particularly taxing mix of high stakes, violence, confusion, and urgency. With networks providing plentiful information, the opportunity exists to enhance informed reasoning and analysis under such conditions, time permitting. Yet because time often does not permit, it is just as important to strengthen intuition, even while being aware of the biases of one's mental models and the limits of one's experience. The key, as shall be shown, is to integrate intuition and reasoning—the yin and yang of cognition.

We are not the first to flag the heightened importance of reasoning and intuition in the context of networked warfare. The issue has generated significant interest as well as a growing body of good and timely research. However, instead of treating cognition as a detail to be worked out so that information networks can fulfill their promise—as some have done—we see it the other way around: networking offers a golden opportunity to improve the power of thinking under fire.

The mind is often inadequate to solve complex problems, such as those that will be commonplace in the unpredictable military contingencies of a fluid future. But the mind is gifted beyond any machine—indeed, beyond its own comprehension—and it now has high-performance, distributed information systems to assist it. Far from being less important, reasoning and intuition are more important, and they hold new potential. In any case, computers and networks cannot be held responsible, but people can.

The role of the mind in networked warfare is still unknown. This book is meant to raise ideas, issues, and possibilities, as well as—at the risk of seeming presumptuous—a potential framework. Some new concepts and terms are introduced, and some established ones are heavily used. To assist both our explanation and the reader's understanding, we provide a short glossary of key terms at the back of the volume.

This book is written for the policymaker, the strategist, the warfighter, and the layman. At the same time, we hope that the research community will benefit from our attempt to put battlefield cognition into strategic and policy contexts. Clearly, more empirical and theoretical research is needed, which is why we conclude the book with a set of questions for further analysis. Where we have suggestions to make, they are only indicative and, we hope, provocative. As this field continues to evolve, we can be sure of only one thing—ours is far from the last word.

Acknowledgments

A long the journey leading to this book, we acquired a huge debt to our colleague, Courtney Richardson—a superb researcher, thoughtful critic, and relentless trail-boss. We thank Martin Libicki, Dave Signori, Harry Thie, Hans Binnendijk, and Jim Hosek for giving us the benefit of their insights, experience, and advice. The understanding of our significant others—Cynthia Gompert, Stefanie Schmidt, and Ayari De la Rosa—has been as indispensable as ever. Thanks also to the Office of the Chief Information Officer of the Department of Defense for its support. Lastly, the Center for Technology and National Security Policy and the National Defense University, of which the Center is a part, deserve special recognition and gratitude for providing the time, resources, and encouragement for such a speculative piece.

Battle-Wise

Chapter One

From Firepower to Information Power to Brainpower

Approaching a New Threshold

From cornering terrorists to stabilizing war-torn countries to waging all-out war, the courses and outcomes of military campaigns are increasingly shaped by networks that enable dispersed units to collaborate by sharing data. Along with the high-precision sensors and weapons they connect, networks are turning information power into military power.[1] In the logic of the U.S. Department of Defense (DOD) Office of Force Transformation, networks permit information-sharing, information-sharing enhances shared awareness, and shared awareness enables collaboration and speed—the keys to military effectiveness.[2] Defense investment priorities are shifting from mechanized platforms and weapons to the information collectors, processors, links, software, and services that compose these networks.

With its unmatched defense resources and technological talents, the United States has pioneered networked warfare, as it pioneered information networking in general. But from now on, the United States will have company, both friendly and not. For example, China and al Qaeda, using different doctrines for different purposes, are showing interest in tapping the power of information. The Chinese face high cultural and organizational hurdles on the path to realizing the military potential of "informationalization" (as they call it). However, al Qaeda and its affiliates already are showing ingenuity and resourcefulness in putting networking to work with virtually no investment.[3]

As adversaries start to exploit networking, the United States must seek new leverage by improving its warfighters' ability to make sense and use of information in war's confusing, severe, and violent conditions.[4] While no amount of information, no matter how good and timely it is, can remove the strange and ambiguous circumstances of war, the mental faculties of military decisionmakers—from lieutenants to lieutenant generals—are more crucial than ever. The next defining capability in the evolution of warfare, and arguably the highest plane of strategic competition, will be that of the soldier's mind.

Yet little programmatic emphasis is being given to understanding, let alone developing, the particular human cognitive abilities that matter most in making good decisions during networked warfare. On the Pentagon's current list of seven key elements for implementing what it calls "network-centric warfare" (NCW), improving the quality and speed of operational reasoning and problem-solving by people so that they can take advantage of networks has no place.[5] None of the "four pillars of transformation" of the U.S. Armed Forces has to do with adapting human decisionmaking to match either the military-network revolution or the turmoil in global security.[6] Official documents note the growing significance of the "cognitive domain," but no coherent strategy to excel in that domain has been forged. Even the best of the literature on military networking provides little insight into how thinking is affected and can be improved.[7] Students in military training and education programs are being taught how to manage networks and manipulate data, as they should be, but they are not necessarily learning how to reason within the new networked environment, which is surely as important.

Networking in Warfare: End or Beginning?

Networking is the most recent leap in how humans fight. Since warfare began, every favorite weapon eventually has been outdone by a better one.[8] Clubs could defeat fists but were in turn defeated by spears and arrows, which then gave way to guns and bombs. With industrialization, placement of guns and bombs on mechanized vehicles, such as tanks, submarines, and airplanes, brought decisive advantages in mobility, range, survivability, and explosive force. Missiles and sensors then increased the range, speed, accuracy, and lethality of explosive force. Throughout, those who mastered the production and use of each new technology have held at least a temporary strategic edge over stragglers, as Britain did on the high seas in the 19th century, Germany on land by the outset of World War II, and the United States by the end of World War II and again at the end of the Cold War.

We have reached the point where forces with information links among mechanized platforms, high-precision weapons, advanced intelligence collectors, well-trained fighters, and efficient command centers can make quick work of modern but non-networked mechanized forces. The swift trouncing of Iraqi army and Republican Guard divisions by smaller but networked American and British ground, air, and surveillance forces in Operation *Iraqi Freedom* marked the passing of the age of 20th century warfare.[9] Such connectivity, wisely used, also can improve military performance in noncombat operations, such as peacekeeping and humanitarian relief.[10] By enabling any part of a force to operate with any other part, networking affords not just unprecedented capability but unbounded opportunity. It has been called the "apotheosis of conventional warfare."[11]

Strictly speaking, the networking of forces is not new. Britain's Royal Air Force relied on its newly invented radar to direct its interceptors toward incoming waves of Luftwaffe bombers. Anti-submarine warfare has long depended on links and collaboration among surface ships, aircraft, sensors, and hunter-killer submarines. In a precursor to today's joint strike-maneuver operations, the North Atlantic Treaty Organization (NATO) sought to link its air and land forces to toughen defense against Soviet armor. In all these cases, however, communications links were wanting. While the concept of networking forces is not new, the ability to send and receive broad streams of data is—compliments of the fusion of computing and telecommunications that began in the civilian world about a quarter of a century ago. The technologies of distributed computer processing multiply the awareness, speed, combinations, uses, and efficacy of networked military forces.

Information technologies and networking give U.S. forces more and better information, greater weapon accuracy, and a way to disperse platforms while coordinating and concentrating firepower. These technologies can make virtually any vehicle-sized object an illuminated, reachable, and vulnerable target.[12] They increase the probability of destroying that which warrants destruction while sparing that which does not. By improving dramatically the economics of sharing data and providing horizontal (peer-to-peer) links, networking can make forces better informed and able to collaborate than old-fashioned stove-piped ones, within which information flowed vertically and slowly and between which it flowed only at the top, if at all.[13]

Does networking constitute some ultimate stage in military capability and performance—an end of military history (if, sadly, not an end of wars)? After all, since every other imaginable military capability and structure can, in turn, be enhanced by being networked to others, it is hard to think of a better way to form and operate forces.[14] An aircraft carrier is a potent war machine, but two of them operating in tandem are more potent than the sum of two operating independently. No matter how strong an army brigade or an air force squadron is, they are bound to be stronger—indeed, they have been shown to be stronger—when networked together. By enabling forces to be more lethal, agile, precise, fast, elusive, survivable, supportable, and coherent (regardless of distance), networking provides an edge over history's prior accumulation of conventional weapons.[15]

Military exploitation of data networking has progressed swiftly over the past decade or so. At first, data retrieved from remote sensors provided exact locations of fixed targets, which enhanced the accuracy of precision-guided munitions and the effectiveness of strike operations. Before long, linked sensors and shooters were operating as combat teams. Then, the fusion of data from multiple sensors improved the battlefield awareness of individual warfighters and units. Still more value was added when awareness was shared, fulfilling the force commander's dream of a common operating picture of both enemy and friendly forces. All of this occurred roughly during the decade following the 1991 Gulf War (when U.S. military strategists were wise enough not to let victory cause complacency).

Another rung on the ladder was reached when networked sensors and shooters with shared awareness began to collaborate fluidly and effectively in Operations *Enduring Freedom* (2002) and *Iraqi Freedom* (2003). As of now, decisionmakers throughout the force have available unprecedented riches of information and options from which to choose.[16] Throughout this climb in the value of networking, the role of cognition has expanded. On the next rung, cognition will be supreme.

Networking involves much more than merely arranging industrial-age forces in an information-age constellation. Military networking, like networking in other sectors, rewards ingenuity in designing systems, conducting operations, and reforming organizations. For example, relying on the global positioning system (GPS) instead of onboard guidance systems has both improved the accuracy of missiles and reduced their cost. Substituting airpower for the heavy artillery that ground forces have had to lug around has made those forces more transportable and maneuverable, yet every bit as powerful. Global and regional joint commands have been formed to plan, prepare, and conduct integrated operations. Entirely new capabilities, like the U.S. Navy's sea basing, are being created specifically to exploit—and could not succeed without—networking.

Connecting People

"Networks of what?" the reader may ask.[17] Narrowly defined, *networks* are transmission media that interconnect information systems, such as computers, communications transmitters and receivers, and displays of data and images.[18] But these linked information systems form only one layer of military networks. Beneath, and linked through these systems, are the machines and facilities of war: weapons, weapon platforms, sensors, command centers, supply depots, and the like. At yet another level, networks tie together the structural components and echelons (for example, companies, divisions, and strike groups) of forces from the several armed services—vertically, horizontally, diagonally, and adjustably. But most important, if most overlooked, networks connect people—thinking, feeling, responsible, creative, fallible, problem-solving, decisionmaking, sometimes frightened people.

The thesis of this volume is that the most rewarding task, if also the hardest, in seeking to realize the full military promise of networks is to draw and build upon the ability of warfighters to think when linked. The head of the Pentagon Office of Force Transformation got it right when he said that networking will "accelerate our ability to know, to decide, and to act."[19] At the end of the day, knowing, deciding, and acting are the functions—and the responsibility—of people, not information systems.

What makes networking so different from previous military-technological advances is that it can multiply both the information and the choices available to warfighters, assuming they are properly organized and employed to take advantage of it. To be sure, networking does not simplify warfare and may complicate it, or at least make its complexity more apparent. But by supplying more and better information, networking could, if accompanied by improved cognition, produce appreciably better solutions to complex military problems. The countries and groups that excel at converting the potential of networks into faster yet better judgments—better campaign strategies, tactics, and decisions—will be in a position to outperform those that do not. This is already evident in the shrewdness with which terrorists are using the Internet to amplify the fears of the global community and publicize the horror and fanaticism of the disciples of Abu Musab al Zarqawi in the struggle for the future of Iraq.[20]

Networking should not be expected to bring a new equilibrium to military affairs. Nations and groups will scramble to exploit information technology (IT) and network principles, much as advances in propulsion disturbed warfare in the mid–19th century and the introduction of radar triggered efforts to gain an advantage from remote detection in the mid-20th century. As more armed forces and groups make use of networking in the years to come, perplexing questions will arise:

- If two belligerents have networked forces, how can one gain an edge?
- Are the platforms of networked belligerents more vulnerable because networked sensors can see them or less vulnerable because they can be dispersed?
- As more forces gain information power, what is the next defining, decisive strategic capability?
- Of immediate interest, in view of the persistent insurgency in Iraq, how can networked forces use information to defeat a dispersed enemy that is itself networked and hidden in a civilian population?

The answers to these questions, we believe, are right between our ears. The instrument that can make the fullest use and greatest sense out of information—indeed, has biologically evolved to do so—is the brain of homo sapiens. Just as networking is the key to transitioning from firepower to information power, the mind is the key to graduating from information superiority to "time-information" superiority, a concept we will develop in due course. With the right approaches to improving decisionmaking in combat, such superiority is attainable even as enemies adopt networking. The erosion of the American monopoly in harnessing IT for military purposes creates a need to define, gain, and hold a lead in the ability of soldiers to think soundly and quickly while engulfed by confusion and violence. This explains the budding interest in defense-research circles about the cognitive implications of NCW.[21]

Debate over how people think and decide is not restricted to the military sphere. Corporations are growing more concerned with how their employees are using information, as well as how they can recruit and develop people with exceptional cognitive abilities. The information revolution and advances in neuroscience have spawned a new popular literature in human problem-solving and decisionmaking. In *Blink: The Power of Thinking Without Thinking*, Malcolm Gladwell argues that people can arrive at quick yet sound judgments by unconsciously noting those flakes of information that matter most in a blizzard of data.[22] In *The Wisdom of Crowds*, James Surowiecki makes the case that, under many conditions, groups of people make smarter choices than smart individuals do.[23] (Later we will address how these propositions apply to battlefield decisionmaking.) Even as IT is getting faster, smaller, ubiquitous, and more practical, a shift of interest is occurring toward what it all means for human thinking.

Nothing, of course, is novel about the idea that superior thinking can win battles. Brilliant generalship has always mattered, sometimes more than force strength. Lee outfoxed Hooker at Chancellorsville. "Stonewall" Jackson befuddled and beat three Union armies in his Shenandoah Valley campaign. Eisenhower's D-Day plan fooled Hitler's best generals. In Vietnam, Giap got the better of Westmoreland. But networking offers more: an unprecedented opportunity to prevail in battle by bringing to bear more general brainpower—not just brainier generals—by enabling better problem-solving on the part of the individual, by mobilizing the minds of more individuals, and by honing the collective intelligence of teams.

While information networking can help military organizations and personnel in many ways, the concern here is with war and those who wage it. Streamlining military infrastructure, logistics, and peacetime administration, and improving the performance of the people executing these functions, though commendable, are all beyond the scope of this study. The networking of forces for combat and other demanding operations present opportunities that the U.S. military is only beginning to fathom. Yet networks alone will not ensure that the warfighters connected to them will make full use and good sense of information. That will take a purposeful and comprehensive strategy of its own.

Building a Cognitive Strategy

Such a strategy must span geopolitics, technological opportunity, military operations, command and control, and personnel policy. Crafting it will require the analysis of:

- changing demands of solving problems and making decisions in warfare
- emerging threats that make it crucial for U.S. forces to meet those demands
- cognitive abilities that matter most in operations against those threats
- cognitive effectiveness based on lessons from wider research on and experience with networking and decisionmaking.

Following this logic, chapter two examines how the complexity, urgency, tension, and information ebbs and flows of warfare may challenge the cognitive capacity and judgment of military decisionmakers. Chapter three looks at the prospect that U.S. forces will have to face networked opponents, and chapter four derives key cognitive abilities as well as deeper traits that could provide an edge in operations against such opponents. Chapter five suggests concepts and techniques of decisionmaking to harness those key abilities. Chapter six looks at the building blocks of superior cognitive performance—individuals, teams, and command and control. Chapter seven is an excursion into wider research and experience, including how leading corporations and other high-pressure, nonmilitary organizations are trying to put more into and get more out of the minds of their people. Chapter eight analyzes policies that can accentuate and develop key cognitive abilities and improve decisionmaking, taking into account the operational challenges, networking potential, and lessons from nonmilitary domains covered in the preceding chapters. Finally, in chapter nine, specific recommendations and strategic observations are offered for consideration.

Notes

[1] We will use *network* in its broadest sense to include not only information processing systems and communication links but also the platforms, sensors, command centers, and troops that make up a force. While improved precision of individual sensors and weapons, owing to IT, partly accounts for the enhanced effectiveness of a force, their precision is increasingly both enhanced and exploited by networks.

[2] Office of the Secretary of Defense, *Network-Centric Warfare: Creating a Decisive Warfighting Advantage* (Washington, DC: Department of Defense, 2004).

[3] Rohan Gunaratna, *Inside Al Qaeda: Global Network of Terror* (New York: Columbia University Press, 2002).

[4] Throughout this book we will refer to *warfare, combat,* and *military operations* more or less interchangeably. Of course, military operations span a broad continuum, including counterterrorist actions, peacekeeping, and humanitarian relief. We are convinced that both information networking and improved cognition and decisionmaking offer benefits across this continuum. Indeed, operations against nonstates and in semipermissive and permissive conditions may be fraught with ambiguity and confusion, even if not at the levels of danger and urgency found in fighting the armies of other states. The cognitive demands of noncombat operations are different than those of combat but are nonetheless formidable, growing, and in need of the same attention. Thus, the thrust of this book can be applied to combat and noncombat operations alike.

[5] Office of the Secretary of Defense, *Network-Centric Warfare.*

[6] Office of the Secretary of Defense, *Military Transformation: A Strategic Approach* (Washington, DC: Department of Defense, 2003).

[7] For instance, David S. Alberts and Richard E. Hayes, *Power to the Edge: Command and Control in the Information Age* (Vienna, VA: CCRP, 2004), is a superb and seminal book on implications of networking. It touches on but does not explore how "sense-making" and decisionmaking are affected and how training and education should be tailored in light of network-centric capabilities and concepts.

[8] Nowhere is the history of military technology better narrated and explained than in Martin van Creveld's *Technology and War: From 2000 B.C. to the Present* (New York: Free Press, 1991), in which he traces the progression from the stone hand-axe and sharpened flint to computers. Van Creveld did

not foresee the impact of networked warfare. Instead, he expected that "cybernetics" and computers would open the door to the "quantification" of warfare, whereby civilian managers, analysts, and economists would have growing influence over everything from logistics to operations. He also warned of the danger that computers would cause militaries to ignore those factors that cannot be reduced to numbers, such as ferocity and morale.

[9] U.S. and allied forces had many advantages, including superior equipment, firepower, quality, discipline, and doctrine. However, what accounted for the remarkable speed of victory for a comparatively small force was the awareness, integration, and precision afforded by information systems and networks.

[10] David C. Gompert, Hans Pung, Kevin A. O'Brien, and Jeffrey Peterson, *Stretching the Network: Using Transformed Forces in Demanding Contingencies Other Than War* (Santa Monica, CA: RAND, 2004).

[11] James C. Mulvenon et al., *Chinese Response to U.S. Military Transformation and Implications for the Department of Defense* (Santa Monica, CA: RAND, 2006).

[12] Admiral William Owens' measure of success in a 200 × 200-mile area.

[13] Alberts and Hayes.

[14] We recognize that networking may fail if communications links are successfully attacked, leaving at least parts of the networked force worse off—less effective and more vulnerable—than had they been organized in a traditional way. So far, the threat of such attacks seems great enough to invest in countering it but not to inhibit exploitation of networking.

[15] Assuming that nuclear weapons remain viewed as highly disadvantageous to the point of being unusable.

[16] The authors acknowledge the insights of Jeremy Kaplan and the late Vice Admiral Arthur K. Cebrowski, USN (Ret.), in recounting this progression of networking.

[17] See Alberts and Hayes. Networks are physical, informational, cognitive, and social.

[18] These are not all dedicated military information systems; global telecommunications and Internet infrastructures are heavily used.

[19] Office of the Secretary of Defense, *Military Transformation.*

[20] Ariana Eunjung Cha, "From a Virtual Shadow, Messages of Terror," *The Washington Post,* October 2, 2004, A1.

[21] For example, see the Defense Advanced Research Projects Agency, *Improving Warfighter Information Intake Under Stress Project,* available at <http://www.darpa.mil/ipto/programs/augcog/>; and Martin Burke, *Thought Systems and Network Centric Warfare,* available at <http://www.dsto.defence.gov.au/publications/2252/DSTO-RR-0177.pdf>.

[22] Malcolm Gladwell, *Blink: The Power of Thinking Without Thinking* (Boston, MA: Little, Brown and Company, 2005).

[23] James Surowiecki, *The Wisdom of Crowds: Why the Many Are Smarter than the Few and How Collective Wisdom Shapes Business, Economies, Societies, and Nations* (New York: Random House, 2004).

Chapter Two

Cognitive Demands of Networked Warfare

Cognition, Complexity, and Information Networks

Severe cognitive challenges obviously are not confined to warfare or to the present. Almost half a century ago, the economist Herbert Simon stated:

> The capacity of the human mind for formulating and solving complex problems is very small compared with the size of the problem whose solution is required for objectively rational behavior in the real world or even for a reasonable approximation to such objective rationality.[1]

This gulf between limited minds and complex problems is due to the inadequate ability of humans to form mental models to help them discern the intricacies of reality. In particular, our mental models of the causes and effects of *complex and dynamic systems*, such as warfare, macroeconomics, and child raising, are grossly simplified compared to the systems themselves. Yet forming more complex mental models seems beyond our capacity. So humans find themselves trapped in a state of "bounded rationality"—with shortcomings in attention, memory, recall, and information processing that limit the ability to comprehend and thus to make sound rational judgments.[2] Worse, the ability to reason can fail when it can be least afforded, particularly in the face of especially difficult problems.

Bear in mind, though, that Simon's observation about the limits of the human mind predated the information revolution. This begs the question: Does (or can) applied IT compensate for deficiencies in humans, mental models and thus improve their ability to solve complex problems rationally? Without doubt, computers and networking—especially data networking, which both distributes and integrates computing power—have begun to free problem-solving humans from their limited attention, memory, recall, and processing capacity. Is this not, after all, what distributed IT is supposed to do? With electrons and photons doing the hard labor of crunching and passing data, can human neurons now concentrate more on the faculties that distinguish the species: reasoning, wisdom, imagination, and weighing normative values? If so, it follows that information networking ought to let humans formulate and solve complex problems as never before.

If, instead, the advent of networked computing and the availability of useful data are not noticeably improving the ability of people to solve complex problems, maybe it is because we now face—or, thanks to IT (ironically), have a heightened awareness of—a more bewildering world. The complexities of reality, exacerbated by that information age nemesis, information overload, still may be too puzzling even for networked problem-solvers. In that case, the promise of networking is bounded by the rationality of humans.

The truth surely lies somewhere between these alternative conjectures: information networking is unquestionably helping humans solve complex problems, but the inadequacy of human cognition when faced with complex problems limits the ability of networking to yield better results. We find no broad-based empirical proof that the quality of human judgment, military or otherwise, has improved generally and appreciably with the spread of information networks—not thus far, at least. Computers are helpful at solving problems and facilitating decisionmaking of particular sorts, specifically when quantifiable calculations, tons of data, or intricate logic are involved. But they do not ease the burden of making tough judgments involving disparate, competing, and subjective values.

Nor does IT guarantee objectivity, clarity, openness, and good judgment in using information. Glaring examples of misread or misused information are easily found. The failure of intelligence analysts and government officials—smart individuals, by most measures—to heed and share warnings about the threat of a spectacular terrorist operation despite vast data memory, computing, and communications capacity, not to mention reports of highly suspicious activity, has been judged as the worst avoidable lapse in the failure to prevent the 9/11 attacks.[3] The infamous "slam dunk" conclusion that Iraq possessed weapons of mass destruction (WMD), reached by none other than the Director of Central Intelligence, is a reminder that considered judgments may be based on a reading of information that does not reflect reality.

Whether or not IT and networking are producing better thinking and decisions, the ability of humans to solve complex problems is the crux of the challenge of understanding and coping with our confusing world, including the military domain. Any thought that cognition has been made *less* important by information networking surely has been demolished by the effect of the Internet, which is already helping much of humanity make decisions on matters of every sort, but not making these decisions for them. If networking has not so far made people better at solving complex problems, including those of warfare, all the more reason to look more closely at human cognition and not simply at how to construct wider and faster networks.

The Centrality of the Individual

People have a distinct advantage over the machines they were clever enough to invent: the ability to make difficult judgments.[4] The industrial age came about because humans, using science and imagination, invented machines that outperformed human brawn and reduced the need for physically hard labor. Machines did not end human work but rather freed up people to make more use of their fine skills and their minds, both in operating machines and creating better ones. The computer age took off because people are better at designing computational and memory devices than at computing and memorizing. As networks ingest, sort, bank, and move data, people can turn to more cerebral endeavors, where their comparative advantage lies, and thus be more productive.[5]

As the power and uses of information systems rise, the human mind will find higher ground, as it is wont to do. People may not even hold for long an advantage over computers in contingency planning and foreseeing future moves, as evidenced by the success of the IBM Big Blue chess-playing computer against the top human player. But people are unrivalled when it comes to balancing and blending data, ideas, and values; combing and combining operational, technical, and political facts; being cautious or courageous, as called for; sensing intentions; imagining and pretending; exhibiting loyalty and trust; knowing right from wrong; and relating to other people.

In this context, the networked, problem-solving person with access to information is increasingly being recognized as the true center of many complex and dynamic systems. Therefore, understanding how individuals reach decisions, and can be helped to reach better informed and sounder ones, is important in improving the performance of such human-centric systems, of which military action is but one case.[6] The above-average high school graduate is another example. Whether he or she opts to go to college, get a job, or join the Armed Forces hinges on his or her ability, with or without parental input, to sort through a vast array of competing information, make sense of the life implications, and choose among quite different possibilities. Higher education systems, job markets, and military force planning all hinge on the informed reasoning of teenagers. Moreover, these systems are made interdependent by the common individual decisionmaker at their center. The better the individual decisions, the better these systems and set of systems work. The Internet provides pathways to vast amounts of information bearing on these decisions, but 17-year-olds make them.

In the field of health care, better informed individuals—aided in many cases by their own Internet research—can shift market power from the medical, drug, and insurance industries to those with the strongest interest in, and ultimate responsibility for, health: the patients. In these and other such cases, the more knowledgeable the end-user and the better his or her ability to locate and pull useful information to assist in decisions, the better the decisions, the greater the leverage of the individual, *and* the better the performance of the whole system.[7] We now see this dramatically in air travel, where the full-service, higher-cost airlines are being challenged not only by cut-rate ones, but also by Internet-savvy customers seeking the best value.

Nearer the military realm, the individual problem-solver also stands at the center of complex systems for managing emergencies, such as public health scares, terrorist threats, and natural disasters. The first responder, informed by firsthand observation and by information from a network, is often in the best position to decide the critical initial course of action. The policeman, fireman, or emergency room physician usually will rely on some mix of intuition and reasoning, with the former preponderant when time is short and the latter increasingly involved as time permits. In turn, information from that first responder can be made available throughout the entire network, helping others to understand what is happening.

If the first responder is unable to handle the crisis, he or she is usually in the best position to judge what sort of help is needed and how urgently, and then to call for it. In a complex disaster, a fire-fighting team leader in need of hazardous-material and emergency-medical back-up cannot afford to pass the call for help up a departmental chain of command, across to other departmental chains of command, all the while waiting for an answer to come back down. In such cases, linked individuals—peer-to-peer in networking jargon—perform better than arms-length hierarchies linked at the top.

Of course, as we know from the unsatisfactory government response to the flooding of New Orleans caused by Hurricane Katrina, hierarchies—even those chartered to handle emergencies, like the Federal Emergency Management Agency and the Department of Homeland Security—are prone to failure. While the availability of information was hindered somewhat by the damage to communications links on the Gulf Coast, the failure to help those stranded by Katrina was largely the result of poor human cognition. The inability of individual decisionmakers up and down the levels of government to comprehend and formulate solutions to nature's sudden complexity supports Herbert Simon's insight that mental models are weak and often flawed representations of reality.

After decades of preoccupation with the technology, organizations, and processes of complex systems, this growing respect for the role of the informed individual is altering the way such human-centric systems work. More emphasis is being placed on giving the individual the ability to draw upon copious and relevant information and the authority to act on it. By increasing the decisionmaking possibilities and market power of individuals, networking increases the importance of the cognitive ability of individuals to use information to reach sound rational judgments.

There is no reason to think that military operations and forces, with the warfighter at the center, are exempt from the general effects of information, complexity, and human rationality that abound in the larger world. No field of human endeavor has experienced more turmoil of late. In none is there greater need for improved, informed decisionmaking in the face of urgency, uncertainty, and change.

The Effects of the Information and Geopolitical Revolutions

Figure 2–1 depicts how the context of military forces and operations has changed radically in two ways over the past 15 years or so. First, as already noted, technologies spawned by the information revolution now can deliver data of unprecedented volume, quality, and speed to forces that apply network principles in the way they organize and operate. Second, with the end of the East-West stalemate, geopolitical upheaval is causing continuing turbulence and unpredictability at the global, regional, and local levels—in military parlance, at the strategic, operational, and tactical levels. As figure 2–1 suggests, it is uncertain which effect will be stronger in the future: the clarity resulting from more and better information or the complexity caused by increased turmoil. The interplay of these two revolutions is complicated by the fact that plentiful information, unless properly organized, can aggravate rather than ameliorate complexity.

Figure 2–1. Interplay of Two Revolutions

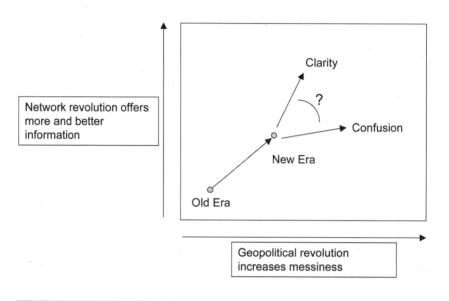

Even before these dual revolutions, making reasoned and timely decisions in the violent crush of warfare was a challenge. The trail of military misjudgments—hopeless head-on assaults, failure to heed clear warnings, unwarranted caution, lost opportunities—rivals the history of mistakes in any field of human activity. The heart of such difficulty is, as Simon observed, that people are not very good at solving complex problems rationally.[8] The more complex the problem, the greater the distance between reality and the model of reality the mind forms to fathom and solve the problem.

If humans are not well equipped to solve complex problems rationally in general, they must be especially handicapped trying to do so during war, when stakes are high, truth is elusive, time is short, and a mortal enemy shares the battlefield. As confusion, intensity, danger, and, above all, urgency increase, military personnel tend to cast aside structured reasoning and rely mainly or exclusively on intuition. Instead of the logical, analytical, inquisitive, and quantitative capacities of the brain, these pressures can cause combatants, and others in analogous situations, to turn to the spatial, imaginative, form-over-details capacities. Thus, even though analysis may be the right response to complexity—analytically speaking—the complexity of warfare may steer the decisionmaker in exactly the opposite direction.

Intuition and Reasoning

It is said that the human brain is the most versatile and finely constructed object known, although its design and functioning are not fully understood. Intuition, the power to attain direct knowledge or cognition without evident rational thought and inference, is one of the least understood aspects of how the brain works. It enables unconscious problem-solving, which may seem simple—like having a hunch or "gut feel"—but in fact involves complex brain activity. Although intuition is more than learning from repetitive experience, it does appear to function more effectively in dealing with familiar circumstances than with strange ones. Indeed, the aspect of intuition most germane to our inquiry into battlefield cognition and decisionmaking is its relationship to prior experience.[9]

Research shows that decisions in combat, as in other intense and urgent circumstances, are made mainly using intuition—the sense of drawing on experience and going with familiar solutions—rather than analyzing and comparing the costs and benefits of multiple options. Up to a point, this is understandable, natural, and desirable. Sound intuition, born of experience, separates competent warfighters from less competent ones and seasoned warfighters from novices.

Any veteran of combat knows it would be wrong to dismiss the importance of intuition and the reliability of snap judgments, even when information is plentiful. In *Blink*, Gladwell details numerous cases in which persons are able to reach conclusions within minutes or even seconds that are as good as those reached through methodical research and deliberation. Such "fast and frugal" thinking—or "thin-slicing"—is based on subtle, even unconscious, rapid screening of information to spot key data that tend to yield sound interpretation and choice. Far from failing in the face of complexity, cognition based on first glance can be of great value when people need to "make sense of a lot of new and confusing information in a very short time."

It is important to note, however, that this ability to judge and decide upon bits of key information, as Gladwell explains it, comes with long-term, repetitive experience, education, expertise, and discipline. Such shortcutting depends on familiarity with the sort of problem being faced.

Moreover, it can lead to mistakes, especially when one does not know whether to trust one's instincts or be wary of them.[10] In essence, it is possible to form reliable models of reality and draw on experience to solve complex problems, provided circumstances are not too new and strange.

Yet it is precisely because complex military-operational problems are often new and strange that they cry out for reasoning, however hard it may be in the circumstances.[11] It can only help, time permitting, to sift through more information, apply logic, and perform analysis, if only to check first impressions and buttress intuition. After all, the main reason to rely on experiential intuition is urgency, not a belief that experience is unfailing or that added data and careful reasoning are superfluous. Yet confidence in experience, and thus intuition, must be tempered by the fact that the conditions and conduct of warfare have become anything but repetitive and familiar.[12] What worked in the Gulf War did not much apply in Bosnia; the Kosovo campaign did not provide a template for Afghanistan; the way Baghdad was taken in 2003 offered few pointers for battling insurgents and terrorists in 2004; and block-to-block fighting in Fallujah will not prepare U.S. forces for a confrontation with the Chinese in the Taiwan Strait. Just as each contingency may be new and strange, so may specific predicaments in which troops find themselves. Being fired upon from the minaret of a mosque is something few if any U.S. Soldiers experienced before they arrived in Iraq.

In the current fluid security environment, the odds are poor that a given soldier heading into a given military contingency will have had enough analogous experience to create a reliable mental model or rely solely on intuition. Of course, experience can be shared, in effect, through training. However, insofar as training is predicated on the set of problems that forces have faced in recent years, its value will be reduced if the next decade is unlike the last one—as it may well be. For this reason, new training methods are needed and are being tried to bolster decisionmaking—despite the unfamiliarity that comes from systemic and situational turbulence.

Because it upsets intuition, change increases the importance of reasoning in warfare. But it hardly makes it easier. The quickening pace of warfare shortens the opportunity for reasoning. Champions of networking claim that speed and the ability to act faster than the opponent are critical to success and that networking is the key enabler of the battlespace awareness necessary for speed.[13] While this is true, networking compresses time in warfare for both sides, not just the other side. In some circumstances, neither enough time nor enough data may be available to evaluate and compare options before acting. So the same forces that make reasoning more crucial also make it harder—too hard to count on it exclusively. The challenge, then, is to improve both reasoning *and* intuition, for both are indispensable.[14] The "co-operation" of reasoning and intuition is as important in warfare as the "co-operation" of the brain's two sides is to effective thinking.

In combat, it may be crucial to think through whether subsequent options are being opened or closed by actions taken, and whether probabilistic outcomes are worth the cost. The combination of unstable security conditions, the predilection for irregular warfare among U.S. adversaries, and the quickening pace of military operations make it at once harder and more crucial to peer beyond the immediate. This puts a premium on the ability to identify not one future but a range of them and alternative ways of approaching or avoiding these futures as well as possible reactions by the adversary. Such complexity historically has challenged human thinking even under calm and peaceful conditions. It explains why persons who need to make hard but prompt decisions rely on intuition instead of analysis, and why they often grab the most obvious solution in their kit of experience instead of pondering the options. Such inherent bias offers all the more reason to strengthen and harmonize both the intuitive and the rational aspects of cognition in combat.

A metaphor for the cooperation between reasoning and intuition can be found in the story of two famous and successful generals, Dwight Eisenhower and George Patton. By all accounts, Eisenhower was the archetypical rational analyst, relying on attention to detail, logic, and objectivity. Patton was nearly the opposite: he was known for his ability to size up a situation with astounding speed, cut to the heart of a problem, and grasp the right course of action without weighing options. While the former strove for perfection in patient planning and consideration of multiple contingencies, the latter moved rapidly from problem to decision to action. According to historian Stephen Ambrose:

> Patton . . . was given to . . . flashes of brilliant insight. [H]e was much taken by his own *déjà vu* and the sensation of having been somewhere before; he devoutly believed that he had fought with Alexander the Great and with Napoleon. . . . Eisenhower had a steady, orderly mind. When he looked at a problem, he would take everything into account, weigh possible alternatives, and deliberately decide on a course of action. Patton seldom arrived at a solution through an intellectual

process; rather, he *felt* that this or that was what he should do, and he did it.[15]

By knowing and valuing each other's character—actually, the thoughtful Eisenhower regarded the intuitive Patton more highly than the other way around—the two generals made a potent duo that proved victorious in one of the most critical campaigns in history: the defeat of the *Wehrmacht* in France following the Normandy invasion.

Does the effectiveness of a strong rational decisionmaker working in tandem with a strong intuitive one mean that these important qualities can be distributed among two or more individuals who are then enabled to work together? If so, it may be easier to realize the cooperation of reasoning and intuition by the way people are teamed up and learn to benefit from one another's cognitive gifts. After all, one of the chief benefits of networking is that it permits individuals of complementary capabilities to combine advantageously. Collective wisdom, operational teaming, and awareness of how colleagues think are all important aspects of cognitive enhancement. However, because the single, accountable decisionmaker is and will remain critical to combat performance, especially when time is short, the military must strive to find, develop, and use the individual—the "Ike Patton," if you will—within whom reliable intuition and strong reasoning coexist and cooperate.

The benefits of an Eisenhower-Patton combination are more apparent when compared with the legendary intuition of another famous figure, George Armstrong Custer. Renowned for his derring-do—unfairly, history paints him as immature, impulsive, not entirely under control, and even "half mad"—Custer was in reality an exceptionally effective intuitive decisionmaker. As a Union cavalry commander, he was "preeminent among his peers" and regarded as a gifted improviser who knew how to "feint, lure, fluster, and tire" his enemy. The so-called Boy General won battle after battle, as well as the confidence of his men. Being "quick in observation and clear in judgment" were especially important attributes for a cavalry officer, given the fluidity and unpredictability of operations.[16]

At the Battle of the Little Big Horn, however, intuition led Custer and his force to disaster. Custer was sure that splitting the 7th Cavalry into a hammer (under his command) and an anvil (under a subordinate) would crush the Sioux force in the middle. In fact, he was so confident that he did not consider what would happen or what his forces could do if the anvil failed to fix the Sioux so that the hammer could fall. While it is unknown, for obvious reasons, what went through Custer's head before the end, historians believe that he and his men lasted no more than 15 panic-filled minutes from the moment the plan failed. One rigorous analysis of the battlefield's artifacts suggests not a valiant "last stand" against an overwhelming force but a risky offensive strategy leading to "sudden, unexpected, and irreversible collapse."[17] Ironically, it was not careful reasoning but failed intuition that left Custer short on time. Yet the lesson of Custer is not as simple as intuition-gone-awry. Rather, it is that exceptional intuition can deliver stunning success in challenging operations, but that the failure to enrich it with objective analysis can deliver just the opposite.

Make no mistake, intuitive decisionmaking can be invaluable in war. With occasional exceptions, such as Eisenhower, great commanders typically are blessed with great intuition. Intuition confers abilities to make quick decisions, size up risks, relate reality to experience, see patterns, sense the enemy's perspective, seize opportunities and initiative, and win the confidence of subordinates. In screening, assigning, developing, and assessing individuals for command, weak intuition should be a red flag.

Our point is not that strong intuition is unnecessary but that it could be insufficient. Again, there are three reasons for this. First, networked information—more, better, faster—can best be exploited in structured cognition and decisionmaking. Second, in today's fluid security and operational conditions, the need for speed, thus for intuition, must be balanced with the need for comprehension of the unfamiliar, thus for reasoning. Third, new methods of rapid-adaptive decisionmaking, like those suggested in this book, allow reasoning to complement intuition without sacrificing timeliness. The ability to use intuitive powers to make "good enough" provisional decisions, to gain both time and information, and to create space for reasoning, despite urgency, offers a way to capitalize on networked information and to confront change. Reasoning should not come at the expense of intuition, nor derogate from its importance. Rather, it can take advantage of good intuition. While the intuitive decisionmaker is still preferred, the intuitive decisionmaker who knows how to blend reasoning without losing time is better still.

The Role of Networking in Decisionmaking

In essence, then, the combination of the information and geopolitical revolutions has:

- compounded the challenge of military-operational problem-solving

- provided tools—those of information networking—to surmount this challenge
- made intuition less reliable but not less important
- increased the demand for reasoning under conditions that do not favor it.

The central question with which these effects leave us is whether networking offers an opportunity for sound and rapid reasoning and, thus, for better decisionmaking in war.

While networking provides many advantages, reducing the demands on military personnel to make reasoned decisions is not one of them. Although IT does a good deal of the mental work previously demanded of humans—navigating, fusing data from multiple sensors, communicating among many and sundry stations, keeping track of forces, targets, and supplies, and computing fire-control solutions—it leaves hard decisions to people. While networked coalition forces quickly dispatched Saddam Hussein's army in pitched battle, they subsequently faced more perplexing choices: Do they attack militants holed up in a Muslim holy site? Do they detain all young men in a neighborhood who look like they could be insurgents? Where will the terrorists strike next? Should an untested battalion of the new Iraqi army be depended on?

The value of well-designed and readily accessible data networking is that it can increase the amount, promptness, reliability, and relevance of information, as well as the possibilities of collaborative reasoning available to the decisionmaker. While mental models may not be improved, networking can augment them by efficiently offering a more faithful, timely, and complete representation of reality, along with more options for action.[18]

In warfare, if anything is as cherished as information, it is time (not counting ammunition, of course). Networked information cannot actually slow the passage of time. But it can enhance a quality we call *time-information*—essentially the product of time and information. In decisionmaking, time can be made more valuable if it is used to gather, evaluate, and exploit information.[19] Conversely, the ready availability of credible and useful information can, in effect, make time more productive, compensate for a lack of it, or, in effect, make it last longer. As noted, the more complex and fluid the world security environment is, and the more unfamiliar military-operational conditions are, the greater the need to augment intuition with strong reasoning. Although the quickening tempo of warfare effectively can compress the time available for reasoning, information networking can decompress time. This may increase the opportunity for, and improve the quality of, reasoning, thus improving operational performance.

The quality of a decision improves as a function of both time and information. Time and information show a strong positive correlation: the more time, the greater the chance to acquire more information; the more information, the more effectively time can be used. Moving from past to present to future, the time available to military decisionmakers is declining, but the information available is increasing. More (or better) information can make up for a lack of time. Up to a point, information networking can compensate for a lack of time in that it can conserve or create time otherwise consumed by chasing, gathering, and sifting through data. Thus, the enhancement of time-information, thanks to networked information, could improve the quality of urgent decisions.

To illustrate, figure 2–2 shows the change in quality of decisionmaking as a function of time-information. An increase in time *or* information (from point A to point B) introduces reason into decisionmaking, which begins to improve *quality*. An increase in time *and* information favors reasoning and *significantly* improves decisionmaking (point C). This dramatic improvement in decisionmaking continues until a point is reached where additional increases in time and information provide only diminishing returns (point D). Many of the sorts of problems warfighters face in the age of networked warfare involve abundant information and scarce time. Networking, though, can help problem-solving by enhancing time-information to permit reasoning despite urgency.

Figure 2–2. Improving Cognition by Enhancing Time-Information

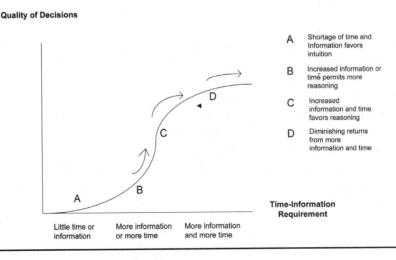

Quality of Decisions

A   Shortage of time and Information favors intuition

B   Increased information or time permits more reasoning

C   Increased information and time favors reasoning

D   Diminishing returns from more information and time

Time-Information Requirement

Little time or information   More information or more time   More information and more time

In addition to helping the individual warfighter solve complex problems and make hard choices, information networking affords warfighters more opportunity to sense, think, and work *collaboratively*. Networking can not only provide collective awareness but also galvanize collective intelligence, or "group wisdom."[20] Moreover, as more individuals are given the chance to use their minds, dependence on a few minds at the top of the pyramid—bounded and fallible, for all their experience and rank—can be reduced. In the future, the term *military genius* may apply less to cerebral admirals and generals than to brilliant forces.

While networking eases some of the cognitive challenges of military operations, it may magnify others. Plentiful information not only can improve situational awareness, but it also can add to the burden of judging the relevance, quality, and accuracy of so much data. By revealing complexity in greater detail, networking can make problems more intimidating. In addition, handling information systems can distract soldiers. Networked information may even lead to the mistaken belief that it spares people the need to make decisions. True, certain decision processes can be largely or entirely automated, such as when a pilot needs only to release a precision-guided weapon upon receipt of unambiguous data from a target-tracking sensor. But the tougher and graver the decision, the more irreplaceable the human. A crucial prerequisite of battle-wisdom is the willingness to take responsibility for solving a problem, as well as for the consequences of the solution.

In the future, military power will depend on the connection between networking and thinking. It is at this nexus that more research, experimentation, and investment should be targeted. IT, the networking that employs the technology, and the information supplied by networking are only tools. They can leave the user better informed, but not smarter. Enhanced time-information only provides an opportunity to reason and thus to make better judgments. It takes the human mind to turn this into a gain in battle.

Messiness, Ambiguity, and Stress

A closer look at the changing international security landscape reinforces the need to move beyond networking in the quest for military advantage. Moving down the ladder of dangers, from a large and advanced strategic challenger to many devious and resourceful sub- and transnational enemies to chaotic civil wars and humanitarian calamities, conditions become both less familiar and less defined for the decisionmakers involved.

The post–Cold War experience, so far, implies a future with many different types of military contingencies, as well as the countless and unforeseeable twists and turns that each contingency can take. Network-enhanced cognition can help people grapple with such variability.

Perhaps the greatest ambiguity is that war and peace are becoming less dichotomous. The boundary between them is both fuzzy and fluid—shifting from one village to the next, from one day to the next, even from one observer to the next on the same day in the same village. Yesterday's bystanders could be tomorrow's enemies or allies. Soldiers must not only distinguish between conditions of war and peace but also make sound decisions in the treacherous gray area between them. This can introduce cognitive demands more taxing than in outright war—demands that fall mainly on soldiers in the field, who are face-to-face with the messiness of reality.

Ambiguity challenges both the intuitive and rational components of cognition. In Iraq, warfare, terrorism, civil unrest, and crime occur simultaneously; failure to tell the difference and act accordingly can cause dreadful results, such as going too easy on terrorists or too hard on unruly but non-threatening citizens. Because military forces are increasingly involved in situations other than war, they must be able to judge whether an apparent opponent is an enemy to be destroyed, a criminal to be apprehended, an angry citizen to be pacified, or perhaps a desperate human to be helped. Doing what is right demands an understanding not only of the efficacy but also of the legitimacy, even the morality, of an action. Networking cannot do this. Warfighters must.

As discussed above, information from networking can save time to enable reasoning in every sort of contingency, from nonpermissive to semipermissive to permissive.[21] But the more chaotic the operating circumstances, the less that information, however well collected, processed, and distributed, can substitute for judgment. To a degree, computers can indicate where and when to shoot an unambiguous enemy. However, the unforeseen contingencies and unfamiliar circumstances in which forces often find themselves in the new security environment demand more and better human judgment.

In general, the urgency, intensity, and ambiguity of warfare militate against considered reasoning in decisionmaking. Three other distinguishing characteristics of war can challenge judgment: the requirement to know right from wrong, the presence of an intelligent enemy with diametrically opposite aims, and the growing interdependence and collaboration among military units that are both distributed and integrated by networking.

More than most human endeavors, warfare is fraught with moral dilemmas and judgments. Our cause is right; the opponent's is wrong. Killing enemy combatants in action is right; killing noncombatants is wrong; killing suspected combatants might be right or wrong. Ending violence is right under some conditions and wrong under others. Aggression is wrong; self-defense is right; preemption may be right if attack is likely and imminent. Courage is admirable; placing one's troops unnecessarily in harm's way is shameful. Ethics not only exceed the capability of any computer, thankfully, but also make rational military decisionmaking that much harder.

Locking horns with a determined and mortal enemy also separates military decisionmaking from other cognitive challenges, except for law enforcement and particularly nasty competitive business markets.[22] The most formidable military opponents are those who are able to confuse and complicate the decisionmaking of one's own forces. An adversary who is trying to seed confusion makes warfare a highly dynamic system, and far more complex than any mental model of reality a military decisionmaker can form.

Interdependence demands that decisions take into account the activities, contributions, and needs of friendly forces, especially those involved in networked collaboration. It means, at a minimum, that warfighters who use the network, not just senior force commanders, must appreciate how their choices and actions will affect others and how the information they have could benefit others. Because networking permits integrated, cross-service operations, these interdependencies may be with units and people that are distant, diverse, and unfamiliar. While decentralization of authority and horizontal collaboration can improve the performance of a networked military force, they also multiply cause-and-effect connections and the potential for unintended and unforeseeable consequences. As networking pervades forces and operations, integration will deepen, interdependence—of information, people, and action—will increase, and decisionmaking will become even harder.

Compared to traditional formations, a networked military unit depends more upon capabilities, such as sensors, supplies, and firepower, that are not under its control or necessarily in the same immediate chain of command or service. In turn, a unit may be depended on just as vitally by others to provide information, support, or reinforcement. The U.S. Special Operations Forces (SOF) operating in Afghanistan were feeding data to Air Force and Navy air planners and pilots and, in turn, relying on the resulting strikes to help them defeat enemy forces.[23] Such links may change, perhaps suddenly and often, in the course of battle.

As if military problems were not hard enough, now they are interwoven. As a unit's actions affect and are affected by others on the network, its decisionmakers must take these dependencies into account, adding to the complexity and pressure of war. The commander of air operations must integrate the use of land- and ship-based aircraft, respond to calls for support from multiple ground units, and take account of targets that can be destroyed by missiles, helicopter gunships, or SOF instead of airpower. Optimization of weapons-on-targets and other resource-allocation tasks can be aided by computers, but the decisions and responsibilities will continue to fall on human shoulders.

This combination of moral dilemmas, intelligent opponents, and interdependence amplifies Simon's warning that the dynamic systems of the real world may be too complex for the human mind to fathom and for mental models to accommodate. It underscores the importance and difficulty of improving both the reliability of intuition and the speed of reasoning.

Granting Authority and Taking Responsibility

These special features of warfare point to yet another heavy demand on the warfighter: the willingness to take responsibility for decisions and be held accountable for the results. Our armed services have long believed and taught that this attribute, not hierarchical standing or career longevity, is a hallmark of true leadership. But the decentralization of decisionmaking authority, made possible by networking and made necessary by complexity, means that more people, and more junior people, will have to take responsibility. Far from making leadership less important, networking gives it a more expansive meaning and demands more of it.

The attitude toward information and authority in the senior ranks is crucial, of course. It is senseless—battle-dumb, if you will—for top commanders to hoard information when networking allows them to share it.[24] But authority can be hoarded, too. When it is, the value of distributed information can be drained by commanders who fail to grant subordinates in the field the authority to decide how to handle threats and opportunities within broad mission guidance. Before networking, commanders could have legitimate concerns about whether juniors had enough information to decide wisely. With networking, any lingering reluctance to diffuse authority presumably stems from the doubts of seniors that juniors are sufficiently trained or experienced to be trusted.

Such doubts are exacerbated by the fact that the messiness of military operations in the new security environment increases the potential for nasty and consequential mistakes, such as friendly fire or noncombatant deaths, any of which could be on Western or Arab television that same day. A good leader would prefer to be able to say that he or she, not a subordinate, made a costly misjudgment. With journalists embedded in combat units, the heightened visibility and sensitivity of what happens in the field militates against delegation of authority—exactly the opposite of what networking permits and conditions demand. The answer to this dilemma is not to deplete the value of networking by centralizing control but to invest in the cognitive, problem-solving skills of subordinates.

Lately, the U.S. armed services have been recruiting more people, officers as well as enlisted, with higher education and giving people more education while they serve.[25] This reflects the increasing sophistication of tasks—especially cognitive tasks—required of them. More than that, it reflects the fact that networking permits and rewards the distribution of responsibilities through and down the ranks, from senior to junior officers and from officers to noncommissioned officers (NCOs). Organizations that conform to networking principles and want to tap the talent of informed and enfranchised people expect more from each person.

Distribution of information and problem-solving capacity will improve the performance of a force only if accompanied by readiness of senior leaders to share authority and of junior leaders to take responsibility. There are indications that many senior U.S. military leaders "get it." For all the difficulties U.S. troops have faced in Iraq, examples abound of junior officers and NCOs being given the latitude to decide how to handle delicate and dangerous situations. Just as important, junior officers and NCOs are readily accepting the responsibility and showing initiative and wisdom. *The New Yorker* reports the case of a lieutenant colonel, faced with a mob of Iraqis "shrieking and frantic with rage," who ordered his troops to "take a knee" and point their weapons to the ground. The situation was defused and a bloodbath averted.[26] The urgency, messiness, and unpredictability inherent in coping with an elusive insurgency, professional terrorists, and a volatile citizenry leave little alternative to authority shared and responsibility taken.

Because distributed information permits distributed decisionmaking, subordinate units should have both increased autonomy and correspondingly increased accountability. Loosely speaking, decisions once made by major generals could be made by majors. This obviously will add to stress and strain on the average major (and perhaps to the anxiety of the average major general). But it also could save precious time—another example of how networking can, in effect, counter urgency and expand the opportunity for reasoning. Happening upon a terrorist stronghold, a company commander—analogous to the emergency first responder described earlier—should be able to call an instant submarine-launched cruise-missile strike, instead of going up his or her Army chain of command through a joint force commander and then back down the Navy chain of command. Failure to distribute authority can defeat the purposes of distributed information and forces and can rob a force of the gains in time-information offered by networking.

Meanwhile, the cognitive demands on the major are changing from carrying out instructions and reporting the results vertically to interacting along horizontal and diagonal axes of a network, with both more decisionmaking authority and greater interdependence in a web of many decisionmakers. Because networked forces are better than hierarchical ones at shifting gears, directions, and configurations, the individual must be able to mix reason and intuition with greater flexibility and speed, not just in reaction to shifting orders but as a semi-autonomous actor in a shifting system. If this all sounds chaotic, imagine using deadly force in the midst of it, while taking fire. Military analysts who opine about the imperative of networked warfare should bear in mind the mental demands it places on the practitioners who are responsible for the consequences of real decisions.

An officer charged with leading an assault on a suspected insurgent hideout must be able to process conflicting intelligence reports from multiple sources as well as options for using his own troops, the availability of back-up firepower, different tactics the terrorists could use, the risks to bystanders, the aim of minimizing his unit's casualties, and the danger of triggering antipathy among the population. This must all be done quickly enough to prevent the insurgents from being tipped off and escaping. Those who can make such decisions—from platoon commander to force commander—can expand the capabilities, performance, and survivability of their forces.

In sum, people in networked warfare must be able to think and willing to decide in strange and violent situations with greater speed, situational and contextual awareness, accountability, and interdependence with others, while being deluged with information of uneven importance and quality—all on top of the familiar pressures surrounding decisionmaking in battle. They must overcome the basic difficulty humans have in grasping and solving complex problems, with IT both helping and complicating that task.

Cracking the Urgency Problem with Battle-Wisdom

The analysis thus far suggests that the cognitive effectiveness of warfighters, forces, and decisionmaking must draw upon both reasoning and intuition, and that it must improve. This is a substantial challenge, given that reasoning requires time and intuition requires experience, both of which may be in short supply. The demands placed by urgency, unfamiliarity, and the importance of the interests at stake on military decisionmaking are depicted in figure 2–3:

- the greater the urgency, the greater the reliance on intuition
- the greater the unfamiliarity, the greater the demand for reasoning
- the more important the stakes, the greater the demand for reasoning.

Of course, any given military-operational problem might fall anywhere in this box. Overall, however, 21st-century warfare regularly confronts its practitioners with problems of high degrees of difficulty in all three respects (the shaded box). This is true now more than ever.

Figure 2–3. Problem Dimensions in 21st-Century Warfare

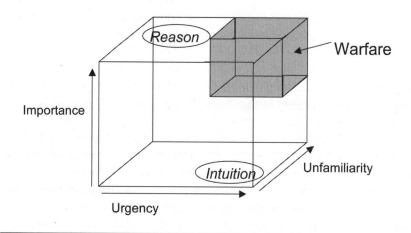

Nothing can be done to make military decisions less important. Similarly, not much can be done to make problems more repetitive and less strange in this turbulent security environment, though some training methods can help people cope with unfamiliarity. However, as already noted, there are ways to alleviate urgency. Doing so would expand the opportunity to reason and reduce reliance on pure intuition. Thus, a central question of this book is how to utilize networked information to ease the urgency problem.

More than merely a combination of timely reasoning and reliable intuition, 21st-century warfare demands the integration of these two components of cognition into the savvy-yet-methodical quality we call *battle-wisdom*. As we develop this concept in the chapters to come, we will see that battle-wisdom embodies particular cognitive abilities—namely, anticipation, decision speed, opportunism, and rapid adaptability—that can make a difference in networked warfare. For now, suffice it to say that battle-wise individuals, teams, and forces are able to create time-information advantages by making swift yet sound decisions using intuition and reasoning in the heat and fog of combat.

The sine qua non of battle-wisdom is that warfighters have access to whatever information they need to have a full and clear picture of reality, inform their decisionmaking, collaborate with fellow forces, and gain advantage over the enemy. Soldiers cannot make more sense and better use of information they do not have. With present technology, the best way to share information through networks is by "post-and-smart-pull," whereby providers make available potentially useful, often new information via the network, and users have the knowledge and technical means to extract and process what is useful for their situational needs.[27] With few exceptions, this is how the World Wide Web works, and why it works so well. Just as the Web defers to inquisitive yet discriminating users, military networking works better when guided by the invisible hand of informed user-need. Among other things, being battle-wise includes knowing what information to pull.

The alternative is for senders—headquarters and intelligence sources—to bombard all users with all information or else to decide specifically what information each user should get. The former is a recipe for information glut; the latter, apart from consuming precious time, overlooks the fact that the user usually knows his or her own needs better than the sender does. As Google, Yahoo, and others have shown, finding information is done by powerful and efficient search engines with, at most, modest regard for the relative utility, refinement, and priority of the data they retrieve. The network software that enables users to search effortlessly and instantly for every morsel of potentially interesting information is arguably the most important new invention since the advent of the Internet.

In contrast, pulling and processing the right information requires a sharp sense of utility and priority (or curiosity plus leisure time)—thus the term *smart pull*. Military headquarters staffs and network managers are not well enough informed to know what each user needs at every point in time. It is more fruitful to improve the ability of each user to know what information he or she needs and can obtain than to try to make headquarters omniscient. In time, the smart user will be aided by "smart-push" technologies and a related facility known as "content staging," which embed in the network knowledge of the user's needs and interests. Even then, the aim of satisfying the warfighter's decisionmaking needs will be paramount.

Potentially helpful information comes not just from headquarters but from anyone on the network. For example, two U.S. Army company commanders (without seeking Army approval or support) created a Web site known as Companycommand.org and, later, one called Platoonleader. org.[28] Although these sites are not part of any combat-operational network, warfighters in Iraq have found them very helpful in coping with strange and hazardous situations. Of course, those who post potentially useful information on such sites are virtually clueless about the specific information needs of any given user at any given moment. The same can be said for operational networks. In this sense, some form of smart pull (preferably assisted by the network) is and will remain fundamental to enabling battle-wisdom.

In addition to what they pull from networks, battle-wise warfighters must be good at sensing local, immediate circumstances, as soldiers have throughout history. After all, there is no reason why networking would reduce the value of that which can be seen, heard, and felt. On the contrary, thanks to networking, first-hand information can be of benefit not only to the individual and his or her unit but also to others and to the whole force, which will be more agile and adaptable if every element is good at sensing and posting. Therefore, each individual has an obligation to report all he or she sees that might be significant so that it may be added to the rest of the network's information to the benefit of all.

At the same time, humans are far from perfect in communicating what they observe. They possess *tacit knowledge*—that which defies articulation and communication to others, and may even be subconscious—or

BATTLE-WISE **35**

they may be too busy or embattled to make timely reporting their highest priority at all times.[29] Because of such tacit knowledge, local personnel are at an advantage over remote ones, including top command, in possessing *both* whatever relevant information is posted on the network (assuming they know to pull it) *and* their latest, grainiest, if subconscious, sense of intermediate circumstances.

To illustrate, the leader of a small unit engaged in fighting an urban insurgency can use all relevant information available on the network and can observe subtle differences between enemy fighters and their rowdy but not dangerous sympathizers—and synthesize these two perspectives. Assuming that no potentially helpful information is withheld from the unit in the field because of security policy or network limitations, no headquarters can match this level of awareness.

One of the benefits of networking forces is that they can be more dispersed, as well as lighter and fleeter, than pre-networking forces, making each unit a potential scout, extending the coverage of surveillance, and gathering more information for others on the network. Yet because not all local information is communicated, there exists some significant amount of "dark information" that is known but is not on the network. Consequently, although the force headquarters may be able to see a *common* operating picture, it can never see a *complete* operating picture. The fluid conditions and wily enemies of the contingencies of today and tomorrow are best handled by local leaders and initiative based on an operating picture that is as complete as can be *at that level.*

The combination of tacit knowledge and information smart-pulled from networks should give the warfighter the ability to use reasoning to augment, check, and take advantage of intuition. Broadly speaking, tacit knowledge may be most important for intuition, while explicit networked information is more valuable in reasoning. It follows that the greater the quantity, quality, and timeliness of networked information, the greater the potential for reasoning—and thus the possibility of meeting the combined demands of urgency, importance, and unfamiliarity that define the warfare box of figure 2–3.

In sum, as table 2–1 shows, the battle-wise warrior can integrate timely reasoning and reliable intuition to make hard decisions and solve complex problems using a combination of local and networked information, despite a lack of time.

This is the basic battle-wise formula for meeting the cognitive demands of operating in an era shaped by dual revolutions. It is the key to unlocking the time-information treasure on which success in networked warfare will depend.

Table 2–1. The Combination of Reasoning and Intuition

Reasoning	Intuition
Needed for importance and complexity	Needed for urgency
Smart pull of information to support analysis	Instant cognition and tacit knowledge
Universal to the network	Unique to the individual
Depends on intelligence and education	Depends on recognition and experience
Ponder	Sense
The individual local to the situation is in the best position to combine reasoning and intuition.	

Notes

[1] Herbert A. Simon, *Models of Bounded Rationality*, vol. I and II (Cambridge, MA: MIT Press, 1962).

[2] John Sterman, *Business Dynamics: Systems Thinking and Modeling for a Computer World* (New York: McGraw-Hill, 2000).

[3] National Commission on Terrorist Attacks Upon the United States, *The 9-11 Commission Report: Final Report of the National Commission on Terrorist Attacks Upon the United States, Authorized Edition* (New York: W. W. Norton, 2004).

[4] It is moot whether this advantage will vanish in the foreseeable future with the advent of artificial intelligence. The information age, so far, suggests that the domain of the human mind is expanded rather than diminished by better, faster, networked computers.

[5] Witness the exceptionally strong and long rise in U.S. labor productivity in the past 10 years as a consequence of investment in IT.

[6] Examples of how conceptualizing the individual at the center of the network is now helping systems analysis include military recruitment, transportation planning, health and insurance planning, welfare reform, criminal justice options, and education policy.

[7] Such information is often unprocessed and unrefined, in the sense that its quality has not been enhanced or manipulated by anyone on the network. This is bad news and good news for the user—arguably more good than bad for the smart user.

[8] Herbert A. Simon, *Administrative Behavior*, 2d edition (New York: Macmillan, 1957).

[9] When we refer to *experience*, we mean training as well as participation in actual operations.

[10] Malcolm Gladwell, *Blink: The Power of Thinking Without Thinking* (Boston: Little, Brown and Company, 2005), 14–15.

[11] *Reasoning* is defined as the power of comprehending, inferring, or thinking, especially in orderly, rational ways.

¹² We do not discount the school of thought that pattern recognition can be valuable. Rather, our point is that when each contingency differs significantly from previous ones and is itself continuously changing, recognition can be severely challenged and cannot be counted upon.

¹³ Office of the Secretary of Defense, *Network-Centric Warfare: Creating a Decisive Warfighting Advantage* (Washington, DC: Department of Defense, 2004).

¹⁴ Another problem with relying on intuition is that it is completely idiosyncratic and thus not repeatable—each person's perceptions, experiences, and thought processes are different. This makes it difficult to rely on intuition for planning or to create a common understanding of a given situation, since each person may see a similar situation differently. For this reason, a more formal decisionmaking process is better suited to developing plans and ensuring that command intent can be understood across multiple actors. This is especially true when one considers multinational or coalition operations.

¹⁵ Stephen E. Ambrose, *Americans at War* (Jackson: University of Mississippi Press, 1997).

¹⁶ Gregory J.W. Urwin, *Custer Victorious: The Little Bighorn Reexamined* (Lincoln: University of Nebraska Press, 1990).

¹⁷ Richard Allan Fox, *Archaeology, History and Custer's Last Battle: The Civil War Battles of General George Armstrong Custer* (Norman: University of Oklahoma Press, 1988).

¹⁸ Because networks also may confuse by providing a glut of information, we must look to improvements in information management to help ensure, through filters, displays, and other techniques, that more data truly informs. In addition, the principle of smart pull should reduce unwanted and unhelpful information. By aligning information management with the demands of the smart user, networks themselves can be made smarter in the sense of being designed and operated to be more discriminating and useful regarding what and how information is made available.

¹⁹ More deeply, the fungibility of time and information could reflect that time is, in essence, the receipt of more, new information. Is time meaningful absent new information? Can new information appear with the passage of time? Of course, time has value other than in providing more information, such as in allowing troops to move across distance to surprise, reinforce, or escape. Nonetheless, it seems that time-information is a valid expression of a quality of increasing importance in warfare.

²⁰ James Surowiecki, *The Wisdom of Crowds: Why the Many are Smarter than the Few and How Collective Wisdom Shapes Business, Economies, Societies, and Nations* (New York: Doubleday, 2004).

²¹ David C. Gompert et al., *Stretching the Network: Using Transformed Forces in Demanding Contingencies Other Than War* (Santa Monica, CA: RAND, 2004).

²² Business can foment fierce competition, and when competitors become dedicated to each other's annihilation they have, in effect, created conflict conditions, making decisionmaking that much more military-like and stressful, but business competition does not approximate the ferocity and stakes of combat.

²³ The ease with which U.S. SOF were able to draw upon and help other forces in Afghanistan, although encouraging, is no cause for complacency over the challenge of integrated joint operations. SOF have been conditioned for a generation or more to work with any and every service and to ignore organizational seams in the way they think and carry out their missions. They offer a good example, but one not easily followed. In Operation *Anaconda*, there were problems with the way SOF movements were not relayed to all battle forces.

²⁴ One legitimate reason for commanders to withhold information is concern for its security. With secure networks, as well as security-conscious junior officers, cases in which information is kept from tactical commanders on security grounds should be rare.

²⁵ For example, the U.S. Navy increased the percentage of new recruits with college experience by 60 percent from 2003 to 2004 and has a goal of 15 percent of recruits with college experience in 2005.

²⁶ Dan Baum, "Battle Lessons: What the Generals Don't Know," *The New Yorker*, January 17, 2005, 42–48.

[27] David S. Alberts and Richard E. Hayes, *Power to the Edge: Command and Control in the Information Age* (Vienna, VA: CCRP, 2004).

[28] Baum.

[29] See Gary Klein, *The Power of Intuition: How to Use Your Gut Feelings to Make Better Decisions at Work* (New York: Random House, 2004).

Chapter Three

Beyond the Networking Advantage

The Networking Advantage

In fits and starts, the United States is transforming its military forces based on networking principles. Despite a burst of political enthusiasm for force transformation following the 2001 al Qaeda attacks on New York and Washington and the ensuing U.S. intervention in Afghanistan, the process continues to be retarded by industrial inertia, sluggish organizational reform, and tenacious parochialism. After all, networking means discarding traditional operating doctrines, canceling investments that do not fit the new paradigm, and demolishing barriers among military services. Moreover, with much of the U.S. Army and Marine Corps bogged down in Iraq, a countertransformation camp has formed, calling for more soldiers and less technology.[1]

Nonetheless, because information networking is so potent, military transformation will prove irresistible, irreversible, and pervasive, as it has proven to be in most nonmilitary enterprises. Although it could take a decade or more before networking completely redefines U.S. forces and operations, we already know from operations in Afghanistan and Iraq that networking existing forces can vastly improve performance. In the coming years, as new forces like the U.S. Army's Future Combat System and the Navy's Sea Basing are specifically designed to exploit networking, the advantages over older forms of mechanized forces will become even more pronounced.

Preliminary evidence and research, while sketchy, indicate that networking can improve the performance of forces along the entire spectrum of military operations, not just in all-out combat. In peacekeeping and nonpermissive humanitarian interventions, networking can give small, light, fast forces awareness, flexibility, precision, and the option to call for rapid help when threatened.[2] Networking allows forces to be tailored for an assortment of contingencies and missions, thus increasing their versatility and the policy options they offer. In time, multilateral forces sponsored by the international community may be able to defeat genocidal campaigns—as the international community failed to do in Rwanda and in Sudan—with less risk of casualties and greater certainty of success.[3]

Strategic Significance of Improved Cognitive Capabilities

While important throughout history, how individuals are prepared, motivated, and organized to think in combat will become matters of high strategy because of information networking. Countries and groups that understand this and invest in improving the cognitive abilities of their forces will have a fighting edge. Al Qaeda has already made such investments in its variant of networked warfare.

Comparing U.S. forces with those of potential enemies on paper is less meaningful than gauging whether they can meet current and future military-operational challenges—challenges that flow from the heavy and fluid demands facing the United States as the chief provider of security in a disorderly world during a turbulent age. While others cannot match America in IT, or in the military platforms, weapons, and sensors being networked, they can exploit the same basic technology and apply the same networking principles to raise the costs and risks to U.S. forces in given missions and situations—perhaps to politically unsupportable levels. Networking can be used in various ways by various enemies and rivals, from terrorists operating in scattered and temporary urban cells to rising powers like China, which can make it more dangerous and difficult for the U.S. fleet to maintain stability in the Western Pacific.

As we write this, the conflict in Iraq reveals vividly the problems posed by adversaries with even the most rudimentary networking. The distributed structure yet coordinated pattern of attacks being carried out by a coalition of religious-fanatical terrorists and Saddamist killers has made the perpetrators difficult to isolate and defeat militarily, especially when they have international links and backers. Neither the terrorists nor the Saddamists have a stationary center of gravity, the destruction of which would incapacitate them. Like more advanced forces, they can disperse yet be coherent and effective—the more dispersed, the less vulnerable and more lethal, dangerous, and effective in terrorizing law-abiding Iraqis. Even networked U.S. forces, awash in information, cannot destroy such webbed organizations. Indeed, the failure to make full, timely sense and use of information handicaps otherwise formidable U.S. forces.

The looming question is this: How will networked forces perform when opposed by networked forces, even technologically inferior ones? Paradoxically, two forces, both of which can observe each other and be dispersed while concentrating their fire, are at once more and less vulnerable to each other. Some speculative analysis suggests that even dispersed forces can become vulnerable once illuminated by networked sensors and targeted by precision weapons of an opposing force, negating to some extent the value of networking forces.[4] If this is so, those who depend on military networking, like U.S. forces, may be unpleasantly surprised unless they excel on the higher cognitive plane.

Apart from the ability of U.S. forces to operate successfully against this or that adversary, the heavy human and fiscal burdens on the United States of serving as the chief provider of global security are becoming painfully clear. The insurmountable military lead of the United States has not spared it from casualties, such as the 2,000 soldiers killed in Iraq, or from hundreds of billions of dollars in war costs, along with doubts and finger-pointing at home. Transforming U.S. forces according to networking principles will not guarantee affordable operational success, especially as unfriendly actors embrace the same technologies and practices. Unless the United States is prepared to forsake its global security role—with grave implications for American and international security—it must put more into and get more out of the cognitive domain of military affairs.

The Monopoly Will Not Last

The military benefits of networking—increased flexibility, responsiveness, precision, deployment speed, maneuvering speed, and survivability—are now widely known and accepted. The chief reservation about relying on networks is that their links or nodes may be vulnerable to electronic attack, potentially leaving forces worse off for having become interdependent. But so far, thanks to information-warfare defenses, the advantages of networking seem to outweigh this peril. In any case, more and more military establishments have begun incorporating such technology. The British, Australian, Canadian, Dutch, Swedish, French, and Finnish militaries, among others, have made networking the leitmotif of force planning and operating doctrine. NATO also has embraced the theme and is developing a Network Enabled Capability via its new Allied Command Transformation program.[5]

Will military networking be a walled, privileged community consisting only of the United States and its technologically elite democratic friends? Not likely. As al Qaeda has shown, any number of other states and even sub-state groups, some hostile, now use and will continue to tailor networking technologies and concepts to their advantage. The recent mutations of both al Qaeda and the Iraqi insurgency indicate that this is already happening. Precisely because information networking is so flexible, powerful, and accessible, U.S. complacency would be negligent.

The two classic conditions for the breakdown of a monopoly are that lucrative returns attract competitors and that barriers to their market entry are low. Where networked warfare is concerned, the first condition clearly exists. If the success of recent U.S. combat operations has not caught the eye of potential adversaries, the extraordinary impact of information networking in other endeavors surely has. The crucial question is whether countries and groups without the technology and resources of the United States and its close friends can break the monopoly and enter the business of networked warfare.

Barriers to entry are lower than before. The rapid and extensive penetration of networked personal computers throughout the world shows that proficient use of information networking does not depend on technological prowess.[6] The spread of the Internet suggests that it is unnecessary to invent, make, operate, or own information systems and networks to use them gainfully and strategically. Companies that are adept at using IT are sometimes ignorant about and uninterested in the details behind the software and hardware they use. Most modern offices abound with talented network users who have no clue how computer networks work, let alone how they are built.

The spread of accessible IT, infrastructure, and services will continue apace. Steadily declining prices, reflecting declining production costs and fierce competition, have sustained an IT buyer's market and sped the diffusion of IT. As standardized information products become commodities and information services become utilities, affordability will remain a nonproblem. Powerful economic incentives, integrated global markets, and software's gossamer quality will defeat any attempt to stop this technology's spread.

At the foundation level, the United States and its affluent, democratic friends have a military-technological lead that should remain insurmountable for the foreseeable future.[7] Their free politico-economic systems, open societies, and vibrant markets excel in all aspects of IT, from scientific discovery to invention to application to use, and this phenomenon is spreading from civilian sectors into the military. While this lead in creation and initial application is generally safe, it does not mean adversaries that reject such freedoms are unable to employ information networks selectively to cause serious military and security difficulties for the United States. Indeed, as China shows, good reasons abound to think they can.[8]

Meanwhile, Western firms that invent IT, far from hoarding it, are actively distributing it to gain competitive advantages from market access and labor-cost reduction. The knowledge of how to build IT hardware and write software is being diffused from North America, Northeast Asia, and Western Europe into Eastern Europe, India, and China. The purchase by the Chinese of the IBM personal computer line of business suggests that important segments of the IT industry will follow in the footsteps of other manufacturing industries attracted by lower labor costs. The creation of new IT, at which the United States and its allies are sure to remain superior, can barely keep up with the spread of its production, let alone its use.

This process is aided by the Internet itself, which attracts and teaches new users. This is a technology that spreads itself by distributing know-how. The brevity between IT discovery and widespread use is unprecedented.[9] Also, the demand for broadband communications within and among multinational firms and markets has required and paid for global network capacity, with hubs and spurs in every continent.[10] Use of IT infrastructure is even harder to control than is the diffusion of the technology.

An entity, be it a household, company, nation-state, or outlawed group, does not have to be large or wealthy to exploit the global information infrastructure effectively. Indeed, smaller, simpler organizations can be more adept because they can more readily modify their structures, operations, and decisionmaking processes. Without need for research, development, or capital investment, and unencumbered by the baggage of extant systems and traditional practices, they are able to exploit networking applications that may have been designed for big customers. Instead of imitating larger leaders, small groups shape and use networks for their own needs. Under such conditions, they can quickly surpass ponderous users in the ability to apply IT and use networks advantageously.

The same rules apply to international actors that have military missions or ambitions. With access to networking technology and infrastructure becoming easier, affordability is not the obstacle it is with ordinary military equipment, like high-performance combat aircraft. The most important factors in exploiting networking are the imagination of leaders, ingenuity of planners, and aptitude of users—talents hardly restricted to the advanced democratic societies that invent the technology. If learning requisite skills requires spending time in the United States or Europe, arrangements for studies can be made.[11] If markets are already proliferating networking technology and the skill to apply it, gaining expertise for those determined to do so for strategic reasons is surprisingly easy and can be done without attracting notice.

The Advantages of Following

As with most technologies, follower-ship in IT can have advantages, whether the intended use is civilian or military. Potential enemies watch how the United States is networking its forces. They read the literature, much of which (like this volume) is unclassified. They check out the Web sites. They know which commercial-off-the-shelf systems and services are adequate for their purposes. They need not venture into the scientific un-known, with all the costs, risks, mistakes, dead-ends, and financial losses inherent in research. In the grand cycles to which technologies tend to conform, being late or absent at the creative front-end does not preclude success as the technology is productized, commercialized, copied, and distributed. Just as Germany turned the tank against its French inventors in 1940, a determined follower can turn information networking against the United States. Because the technology is now proven, the advantages of following are growing. Because it is ubiquitous, followers abound.

Of course, real and potential adversaries of the United States are not all small and technologically backward. It is possible that the United States will find itself facing a large adversary with talent in IT and sufficiently free markets to foster both the creation and application of the technology and networking. The most interesting potential candidate is China.[12] As the Chinese themselves have stated:

> Informationalization has become the key factor in enhancing the warfighting capability of the armed forces. . . . [The Peoples Libera-tion Army will make] full use of various information resources and focus on increasing system interoperability and information-sharing capability.[13]

Of course, it is not necessary for potential adversaries, large or small, to match or mimic the way the United States is using networking to pres-ent military challenges. Unburdened by the global security responsibilities or full-spectrum military requirements of the United States, they can apply networking concepts and capabilities to niche missions. A good example is the way information networking can integrate multiple layers of surface-to-air missile batteries and air-defense radars, making it easier to track and shoot down penetrating aircraft by concentrating on "leakers." Another is the option of synchronizing surface-to-surface missile salvos and other attacks to stun and disrupt the campaign of a superior foe. Either one can cause serious danger to U.S. forces.

The use of networking enables adversaries of varying levels of sophistication to distribute their forces into small, swift units that can swarm and then scatter. Irregular forces involved in insurgencies or terrorism can become harder to defeat when decentralized and dispersed, thanks to communications networks, including the Internet and cellular telephony. At a minimum, the technology can be used to coordinate and publicize, both being critical functions among irregular forces. Such networking in otherwise primitive forces and movements has been used in the past and likely will continue.

A decade ago, the impoverished Chiapas Indians in Southern Mexico relied on a network of sympathetic nongovernmental organizations to reinforce their uprising and call attention to the brutal reaction of government security forces.[14] The rash of kidnappings and televised executions committed by terrorist groups in the Middle East demonstrates the advantages such groups have by networking: they can strike almost anywhere, despite their small numbers; they are hard to find yet able to expose their grisly crimes to the entire world. Many states and international organizations are deterred from joining, or have left, the U.S.-led coalition in Iraq for fear of this murderous web.

American defense planners may be too fixated on the adoption of networking for U.S. forces and operations to notice that others are taking essentially the same path—on smaller scales and with less technological sophistication perhaps, but with cunning, determination, and potentially spectacular results. It is high time to take notice.

Networked Adversaries, Potential and Real

To analyze why, how, and with what effects adversaries can exploit networking, it is helpful to distinguish three cases: mega-states, middle-sized states, and nonstates, such as terrorists groups or drug rings. The first case can be thought of as posing a strategic (global or regional) international challenge, the second a local international challenge, and the third a transnational but potentially global challenge. We will look at China as an example of the first case, Iran the second, and al Qaeda the third, in each case examining how the use of networking could affect the outcomes of military confrontations with the United States and its friends.

Generally speaking, as countries such as China and Iran and terrorist groups such as al Qaeda exploit IT and network their forces and fighters, they will become less vulnerable to and more capable of locating and striking U.S. forces, even though the latter remain much stronger and better networked. While the United States can respond by attacking the computers and networks of its adversaries, the adversaries can gain considerable protection by using generally available infrastructure and services, especially if these are anonymous or undetectable. How and how well such adversaries exploit networking may depend less on their level of technical sophistication and more on their aims, strategies, and resourcefulness. When it comes to warfare, there is no "digital divide" between those that are able to use information networking and those that are not.

China

China has considerable and growing capabilities in information and networking technology, owing largely to foreign technology transfer into the country for the purpose of gaining access to China's cheap manufacturing labor and vast markets. The Beijing regime itself has been ambivalent about the use of IT in the country, fearing it could stir up and spread dissent. However, while the government continues to try to prevent what it considers seditious Web sites, it will not, and largely cannot, block the technology from spilling into and throughout the country.[15] In addition, the Chinese have shown interest in acquiring commercial networks that extend throughout the region.[16] They seem to have made the acquisition and exploitation of IT a matter of national strategy, despite the political pitfalls. While this undoubtedly has economic motivations, they also have shown signs of interest in adapting IT and networking for military use.[17]

Interservice blockages, a culture of deference to hierarchy, and the reluctance to decentralize command and control will impede Chinese military exploitation of IT. However, a new generation of Chinese military officers, like millions of young Chinese businesspeople, understands and will want to harness the power of information.[18] We must assume that the Chinese will apply networking increasingly, if selectively, in operational and force planning. They already are investing in extended-range intelligence, surveillance, and reconnaissance systems, including space-based global positioning systems (their own limited one as well as Europe's Galileo) for navigation and guidance. Since sensor information has to be fed to Chinese forces to be useful, the Chinese will need to create data networks to do this. Indeed, China's investments in extended-range missiles and submarines would make no sense without networking. As the Chinese put it, "the informationalization of missiles and supporting equipment for command, communications, and reconnaissance" will "markedly increase power and efficiency."[19]

One does not need to ascribe aggressive intentions to China to explain "informationalization." In essence, China is not content with the status quo in East Asia. Its goals include the reunification of Taiwan with China and the loosening of U.S. control in the Western Pacific, which is China's gateway to the region and the world. China's intention to develop the capability to take Taiwan, even if only to strengthen its negotiating hand, has focused its military modernization on neutralizing U.S. capabilities to come to Taiwan's rescue. Specifically, the Chinese want to disrupt, degrade, and delay U.S. naval and air forces long enough to take action to "stop the Taiwan independence forces from splitting the country."[20]

Beyond Taiwan, Beijing likely will find increasingly intolerable U.S. policies and forces meant to constrict China's littoral military activities and preserve unrivalled U.S. freedom of action and influence. The Chinese also may feel compelled to stretch their military reach sufficiently to protect sea lines and choke-points in Southeast Asia, through which increasingly vital oil imports are shipped. Relying on the U.S. fleet to police those waters hardly will be acceptable to China in the future.

Because of their strong economy, the Chinese can invest heavily in capabilities to break the American strategic grip. As they do, the United States increasingly will have to turn to networked forces. Networked maritime and air platforms and intelligence sensors will enable the United States to observe and target Chinese forces venturing beyond the mainland, yet they will be dispersed and distant enough to survive Chinese attack. The vitality of East Asia, the tension between U.S. and Chinese aims, and the growing reliance of the United States on networked capabilities to assure its military advantage, make China's response to U.S. force transformation a critical issue.

While the Chinese have several strategic options, including nuclear build-up and irregular warfare (new forms of Mao's "people's war"), turning to military networking must look appealing. Such a strategy would fit with China's growing general interest in IT and would respond in kind to U.S. force transformation. The Chinese will do it in their own way ("with Chinese characteristics"), with the intent to create capabilities for "active defense" via "leapfrog development" and "going with the tide of the world's military development" toward "informationalization."[21]

To paraphrase a recent RAND study on China's military networking options: the Chinese version of network-centric warfare is likely to reflect China's emphases on operational security, operational control, the stratagem of surprise, and massed rocketry. China is unlikely to try to duplicate U.S. air power or develop a doctrine of highly decentralized operations. The major operational challenge for China vis-à-vis the United States is defeating U.S. strike power, notably by finding and targeting aircraft carriers. Networked sensors and weapons may be one way to solve this problem, though China's weakness in systems integration and reluctance to loosen control could stand in the way. Regardless of whether the Chinese embark on a wholesale transformation of their forces and operating concepts around networking principles, it is likely that they will enhance their investment in extended-range sensors and precision weapons.[22]

In turn, the United States must be prepared to deter and if need be defeat Chinese forces that employ IT and networking principles. This, as shall be shown, will depend increasingly on superior thinking and decisionmaking by warfighters.

The importance of people—how they think and decide, how they are prepared, and how they are organized—has not been lost on the Chinese. They intend to compress command chains, reduce staffs, and stream-line structures. Training will be interdisciplinary and joint, and information-alization will be achieved "by leaps and bounds" by "valuing talented personnel." The Chinese have adopted a "Strategic Project for Talented People" aimed at building a "contingent of command officers capable of directing informationalized wars and of building informationalized armed forces."[23] (It is worth noting that no such project is to be found in the U.S. military establishment!)

Iran

A potential local aggressor, such as Iran, is unlikely to embrace military networking as comprehensively as China. Nor can it begin to match China's sophistication in information networking, let alone America's. Then again, Iran would not need to do so because its military strategy vis-à-vis the United States would be quite different and more limited than China's. Whereas the United States would not likely intervene on the Chinese mainland in a war with China, it does seek to maintain the ability to intervene on the ground against lesser adversaries as a way of policing or altering their behavior, if not their regimes. Consequently, U.S. forces must be able to engage in land-expeditionary operations, including occupation, while also maintaining the skills needed, as in the case of China, to defend against an opponent's military initiatives by sea and air.

For an adversary like Iran, being on the defensive and trying to deny U.S. forces a low-cost conquest translate into lower standards—easier requirements—in military networking. Iran could raise the risks to a U.S. intervention significantly by networking its forces, integrating its otherwise leaky air defense, dispersing its ground forces, swarming against U.S. naval assets in the Persian Gulf, improving its awareness of the location and movements of U.S. forces, and targeting its missiles against U.S. forces for maximum harm.

Conceit or ignorance could cause the United States to assume that the Iranians are incapable of devising and utilizing ways to link their forces using largely commercial data networking products and services. Given their relatively simple needs, the Iranians could exploit available information infrastructure and services without having to invest nearly as much as they have invested in weapons-grade nuclear fuel. Because they may not be able to field large numbers of sophisticated sensors and precision weapons, they may not be able to make U.S. forces much more vulnerable than they are today. But they can spread out and connect their own forces, making it harder for the United States to defeat them. Networking of Iranian forces could turn a quick and decisive U.S. operation into a long, costly, and uncertain one.

Should the United States ever put forces on the ground inside Iran to effect and enforce regime-change, as it did in Afghanistan and Iraq, the Iranians could wage dispersed yet coordinated irregular warfare much more effectively than the rag-tag Iraqi insurgent forces that U.S. forces have had such difficulty defeating. Finally, with their penchant for supporting international terrorism, the Iranians could exploit the global network infrastructure to direct retaliatory attacks on U.S. forces, allies, and interests outside Iran or on U.S. soil. U.S. intervention against Iran is already a chancy proposition, at best; Iranian use of even technologically unsophisticated networking could make the costs and risks prohibitively high—again, with strategic implications in one of the world's most vital locations.

Al Qaeda

Al Qaeda differs markedly from traditional terrorist organizations and is prototypical of the new terrorism. "It is neither a single group nor a coalition of groups; it comprises . . . core bases . . . and satellite cells worldwide, a conglomerate of Islamist political parties, and other largely independent terrorist groups that it draws on for offensive actions and other responsibilities."[24] Al Qaeda has no tanks, fighter aircraft, air defense, or frigates—no forces to be destroyed the way Saddam Hussein's were and China's could be. Nor does it have overhead sensors capable of tracking U.S. forces. It represents a new type of strategic threat—one that taxes U.S. capabilities and cognitive performance.

Since U.S. forces pulverized its strongholds in Afghanistan and Osama bin Laden escaped from Tora Bora, al Qaeda has shunned reliance on large, fixed concentrations. It relies instead on cells and networks of people. And people are extremely hard to find and positively identify without using other people as infiltrators or informants, which is not easy against wary terrorists.[25] While finding terrorists in remote areas is different from locating them in urban neighborhoods, both present difficult challenges. Since they disperse both to strike and to survive, international terrorists must use networks to operate. Thus, while U.S. forces saunter toward network-centric warfare, al Qaeda already has implemented a very different but effective form of distributed violence.

It appears from the post-9/11 activities of al Qaeda and its affiliates, including their operations in occupied Iraq, that they are successfully using U.S. networks to suit their strategies. Networking enables al Qaeda to be more fluid, resourceful, elusive, and flexible. Al Qaeda is to low-capital networking what the U.S. military is to high-capital networking. Its "global information grid" is the Internet. And al Qaeda may be culturally and institutionally more able than the U.S. military, with its ponderous procedures for allocating resources and acquiring new capabilities, to adapt swiftly to changes in the forces and obstacles arrayed against them.

Al Qaeda takes a sophisticated strategic approach to information. "[Its] use of the Internet and videotapes demonstrates that 'perception management' is central to the conduct of its war with the West."[26] It does not require dedicated information-network infrastructure or expensive custom services. Al Qaeda makes use of the Internet for propaganda, recruitment, and training—its own version of distance learning—as well as fundraising, communications, and targeting.

> Al Qaeda's use of the Internet through web sites, email, message boards, and chat rooms allows dispersed members to stay in touch constantly, while maintaining the operational security and compartmentalization demanded by their work, under cover of the Internet's anonymity. . . . Islamic jihad groups also use the Internet to dissemi-

nate training materials, either for those who cannot attend the training camps, or to get new recruits and sympathizers excited about what they could learn.[27]

Being both widely dispersed and relentlessly hunted, al Qaeda takes communications security very seriously and trains its operatives accordingly.[28] It is careful to use nondetectable electronic and human forms of communication. For example, al Qaeda used an anonymous commercial fax network for command and control in the late 1990s.[29] Recently, as the Internet and other public systems have become key components of al Qaeda's global network, encrypted and anonymous communications have become easier. Al Qaeda agents use encryption to communicate from Internet cafes.[30] All this makes detection of and eavesdropping on their conversations difficult.[31] It enables al Qaeda to operate as a highly distributed organization, reducing its vulnerability to counterterrorist attacks, especially to a single knock-out punch. The Internet helps al Qaeda both hide and kill—masking the identities and changing the electronic and physical locations of its people while improving their effectiveness.[32]

Unless al Qaeda terrorists are caught in large concentrations, as they were in Afghanistan, they cannot be eliminated by conventional military means and methods. Indeed, military combat operations against al Qaeda may be increasingly rare because the group has become more dispersed. For the United States and its counterterrorism partners, success will require sophisticated measures and networking of exceptional intelligence and investigative capabilities, infiltration, speed, stealth, and skilled police or Special Operations Forces to track, apprehend, and eliminate them. Even then, al Qaeda's structure is so slippery that it takes a combination of good luck and very specific real-time information to eliminate its members even in small numbers.

Al Qaeda knows that people are its most valuable assets, and it targets its resources at recruiting people with technological expertise and aptitude who are then given internal training or sent to public schools, often for education in computer science, engineering, and electronics. "Recruitment and training for high-tech assignments [are] done very carefully, similar to how a military organization would assess both the intelligence and physical condition of volunteers for special operations units."[33] In addition to its own stable of information experts, al Qaeda can access the talents of sympathizers anywhere.

Networking provides a way for al Qaeda and other terrorist groups, as well as states, to form temporary alliances of convenience. Shifting connections among Abu Musab al Zarqawi's splinter group, other al Qaeda affiliates, Ansar Al Islam, Saddamist killers, and at times even the Shi'ite militants of the Mahdi Army—some of them ordinarily mortal enemies— have contributed to the violence in Iraq and the difficulty of stopping it. Indeed, networking permits such organizational and operational flux among groups that it is not clear at any moment who, what, and where the enemy is. *The Washingon Post,* referring to Abu Maysara al Iraqi, electronic spokesman for Zarqawi, stated, "he's a master at being everywhere and nowhere . . . There's no way of stopping [al Qaeda] anymore."[34]

Al Qaeda is not adopting networking as traditional military forces do. While the evidence is sketchy, it seems that al Qaeda has not yet evolved into the sort of peer-to-peer networking that facilitates horizontal collaboration and synchronized attacks without central command. Rather, its form has been cellular, under central direction, and with the cells not connected with or necessarily aware of each other. However, as it metastasizes and its original leaders loosen control, al Qaeda could become a distributed, fluid, and self-organizing ideological mass of planners, fighters, financiers, and propagandists—some networked, some not—under one brand name.[35]

Al Qaeda is meticulous in collecting information, alert to opportunity, and shrewd in the timing of its strikes. What it lacks in physical capabilities it makes up with awareness, analysis, patience, quickness, learning, and adaptability. Al Qaeda and its agents seem to possess and stress the cognitive abilities that make a difference in networked warfare, including a mix of intuition and reasoning. Precisely because al Qaeda and the forces fighting it are so asymmetrical, al Qaeda must use and enlarge its time-information edge to survive while threatening the survival of its targets. Any counterterrorism strategy of the United States and its partners that ignores the nexus of terror, networking, and cognition will fail.

The al Qaeda threat, while critical in its own right, spotlights a general point: the shrewdness, focus, and determination with which a state or group exploits networking are as important as technical infrastructure and scientific depth. The key to exploiting networking is to develop and empower human beings to solve complex problems—or, in the case of terrorists, to create complex problems. It matters less how adversaries measure against U.S. networking concepts and capabilities than whether they are becoming harder to defeat.

This look at how various types of U.S. adversaries might exploit networking suggests the following conclusions:

■ Networked forces with advanced sensors, weapons, and information systems could be within China's reach over the coming decade, though organizational and doctrinal obstacles must be overcome. While still categorically superior, U.S. networked forces could become less effective and more vulnerable against such forces than against China's current non-networked forces. This could raise the costs and risks of protecting Taiwan and defending against Chinese power-projection in the Western Pacific and cause strategic instability.

■ Middle-sized states will be able to use networking selectively for both regular and irregular forces. Even without advanced sensors and precision weapons, the forces could be made harder to destroy, thus denying the United States the option of low-cost intervention. U.S. failure to defeat terrorists and insurgents in Iraq reveals the problems and pain of post-intervention occupation.

■ Determined or desperate nonstate actors will find it possible, advantageous, and even imperative to network and thus to avoid vulnerable centers of gravity while improving lethality. Such groups network people, making them harder to target and more dangerous. Because of its focus on people, al Qaeda is arguably already further into and moving faster up its learning curve of networked operations than U.S. forces are along theirs.

U.S. Strategic Options for Responding to Networked Adversaries

These conclusions are unsettling, and their implication is clear: U.S. forces must be prepared for a wide variety of adversaries that exploit IT and networking principles in different ways. The United States cannot assume that networking its forces, though necessary, will assure enduring, decisive operational advantages. The loss of monopoly in networked warfare could drive up the costs and difficulties of U.S. armed intervention, on which American security interests and responsibilities could depend. The United States will therefore have to be ambitious, creative, and flexible in how it networks and uses its forces. Even then, the networking of enemy forces, especially their ability to disperse and to find and strike American targets, will demand that U.S. strategy look at options based on but beyond networking.

To say that the United States could be disadvantaged as potential adversaries network their forces is not to say that pursuit of superiority on the cognitive plane is the *only* path U.S. forces can take. To retain its operational and strategic advantages, the United States has several available options. It could:

- improve the volume and discrimination of intelligence-gathering capabilities with more, sharper, stealthier, and tinier sensors
- intensify development of information-warfare (IW) tools to attack computers and communications links to cripple enemy forces that rely on networking
- open up a new front for waging war with space-based weaponry to attack sensing and communications satellites or earthly targets
- build better weapons and platforms to increase the scale and precision of the destruction they can deliver.

It is not our intent to reject such alternatives; indeed, some or all may be necessary. But as strategic options, they all have shortcomings. Improving sensors is certainly commendable, especially for finding, tracking, and targeting enemy weapons platforms. However, the economics of detection do not always favor sensors over their targets. The ability of adversaries to proliferate and disperse ever-smaller platforms, thanks to networks, could lead to diminishing returns on investment in sensors. Indeed, barring some breakthrough, perhaps in biometrics, investment in sensors is already yielding poor returns when it comes to finding particular *people* dispersed among other people—as is common for terrorists and other irregular forces—instead of finding *things*.

It is not clear how U.S. security interests will fare if military competition and hostilities are introduced into space or cyberspace. Full militarization of space has not yet occurred. Ultimately, further development could harm those who rely most heavily on space, such as the United States. Militarization of space also raises philosophical questions about whether that domain should be preserved as a peaceful and available realm for all humanity.

Similarly, warfare in cyberspace may seem a tempting course if one ignores that open, advanced, democratic societies depend most on information systems and thus are most vulnerable to attacks on them. Offensive IW will almost certainly be part of the U.S. strategy to neutralize and defeat networked adversaries. But it will not suffice and could backfire. At most, IW could be developed in tandem with, or as an aspect of, the effort to gain and use network-based cognitive advantages.

Lastly, the United States can, should, and undoubtedly will continue to invest in transformational military capabilities, such as weapons accuracy and range, platform speed and survivability, and transport range and flexibility. It will develop capable vertical/short-takeoff-and-landing aircraft, as well as drones for land, sea, air, and undersea surveillance, and strike. It will exploit networking to enable ground forces to become smaller, lighter, speedier, and more easily deployed, and air forces to hit moving targets. Its sensors will become smaller and capable of loitering and penetrating concealment. And, of course, its networks will become faster, more integrated, more secure, and global, as foreshadowed by the Pentagon's development of the Global Information Grid.

This book's thrust should not be interpreted as opposition to such investments. However, one of its core arguments is that transformation, as just described, will prove to be insufficient, given the demands on U.S. forces and the move toward networking among the forces they may face. If the United States had modest international security responsibilities, it could make do by protecting its commanding lead in networked platforms, precision weapons, and high-performance sensors. But, as noted, the challenge for the United States, as the chief provider of global security, is to retain decisive operational military advantages, not just superiority in capabilities. With the arrival of networked warfare, that will take investments in battle-wisdom.

Investments in proven technologies tend to yield *diminishing* returns. Partly this is because, in general, returns are highest when leverage is greatest relative to conventional ways of doing things, and in part because competitors eventually emulate the investment or craft countermeasures. For instance, greater weapon accuracy may not make enough of a difference to justify the cost of improving upon existing precision weapons.[36] As increments of added operational capability per dollar spent get smaller and the breakthroughs of yesterday become the commodities of today, it is time to look for ways to unlock value at a higher level.

Thus, use of networking by adversaries, potential and real, from China to al Qaeda, demands a search for *increasing* returns on investment. The point of departure for the quest for increasing returns is networking itself—not the machines that are linked by it but the people who conceive, form, and use it to gain awareness and collaborate. Taking a page from al Qaeda, and with an eye on China, the United States must give greater attention to the thinking and problem-solving by which its people exploit

IT in strategy and operations. Investment in minds can pay handsomely.[37] The fullness of history shows that brainpower pays increasing returns on investment, in part because of the fertility of human inventiveness. This is truer than ever in the age of information, and it is truer than ever in warfare.

In sum, if networking makes it possible to improve the thinking and decisionmaking of U.S. forces, and the messy new landscape makes it important to do so, the growing exploitation of networking by opposing forces makes it imperative.

Notes

[1] Tom Donnelly, "Rumsfeld the Radical," *The Weekly Standard* 7, no. 48 (September 9, 2002).

[2] David C. Gompert et al., *Stretching the Network: Using Transformed Forces in Demanding Contingencies Other Than War* (Santa Monica, CA: RAND, 2004).

[3] Clifford H. Bernath and David C. Gompert, *The Power to Protect—Using New Military Capabilities to Stop Mass Killings* (Washington, DC: Refugees International, 2003).

[4] James C. Mulvenon et al., *Chinese Response to U.S. Military Transformation and Implications for the Department of Defense* (Santa Monica, CA: RAND, 2006).

[5] North Atlantic Treaty Organization, *Information Superiority & Network-Enabled Capability* (Norfolk, VA: Supreme Allied Commander Transformation Public Information Office, 2004).

[6] For more details, see Internet World Stats Web site, available at <www.internetworldstats.com/stats.html>.

[7] David C. Gompert, *Right Makes Might: Freedom and Power in the Information Age*, McNair Paper 59 (Washington, DC: National Defense University Press, 1998).

[8] Michael S. Chase and James C. Mulvenon, *You've Got Dissent: Chinese Dissident Use of the Internet and Beijing's Counter-Strategies* (Santa Monica, CA: RAND, 2002), and Robert C. Fonow, *Beyond the Mainland: Chinese Telecommunications Expansion*, Defense Horizons 29 (Washington, DC: National Defense University Press, July 2003). Both titles indicate that the Chinese are able to exploit IT despite a lack of openness.

[9] According to Gilder's Law, bandwidth capacity doubles every 12 months; this is significantly faster than the 18 months it takes to double processor speed (Moore's Law).

[10] Martin C. Libicki, *Who Runs What in the Global Information Grid: Ways to Share Local and Global Responsibility* (Santa Monica, CA: RAND, 2000).

[11] The ease with which foreign students can enter and stay in the United States is being reduced by stricter visa requirements and enforcement in response to the al Qaeda threat. Still, it would seem that the chance of stopping people from studying IT in the United States or elsewhere is about as poor as the chance of stopping the technology's global spread.

[12] India, the other rising star in IT, is less likely to challenge U.S. interests or the international status quo.

[13] State Council Information Office, *China's National Defense in 2004* (Beijing: State Council Information Office Press, 2004).

[14] John Arquilla and David Ronfeldt, eds., *In Athena's Camp: Preparing for Conflict in the Information Age* (Santa Monica, CA: RAND, 1997).

[15] Chase and Mulvenon.

[16] Fonow.

[17] Mulvenon et al.

[18] Ibid.

[19] State Council Information Office.

[20] Ibid.

[21] Ibid.

[22] Mulvenon et al.

[23] State Council Information Office.

[24] Rohan Gunaratna, *Inside Al Qaeda: Global Network of Terror* (New York: Columbia University Press, 2002), 54.

[25] George Friedman, *America's Secret War: Inside the Hidden Worldwide Struggle between America and its Enemies* (New York: Doubleday, 2004).

[26] The Institute for Security Technology Studies, *Examining the Cyber Capabilities of Islamic Terrorist Groups*, March 2004, slide 47, available at <https://www.ists.dartmouth.edu/TAG/ITB/ITB_032004.pdf>.

[27] Ibid., slides 10, 11, 18.

[28] Gunaratna, 80.

[29] Ibid., 76.

[30] Hamas is known to have used email and chat rooms in the United States to coordinate attacks in the Middle East.

[31] Dan Verton, *Black Ice: The Invisible Threat of Cyber-Terrorism* (New York: McGraw-Hill/Osborne, 2003).

[32] Ariana Eunjung Cha, "From a Virtual Shadow, Messages of Terror," *The Washington Post*, October 2, 2004, A1.

[33] Verton, 86.

[34] Cha.

[35] Lawrence Wright, "The Terror Web," *The New Yorker*, August 2, 2004, available at <www.newyorker.com/fact/content/?040802fa_fact>.

[36] The accuracy achieved by U.S. precision-guided weapons during Operation *Iraqi Freedom* was good enough to destroy all fixed targets of interest with such precision that buildings across the street were unscathed.

[37] This is why some companies treat training and education as an investment, no less than research and development.

Chapter Four

Defeating Networked Adversaries

Creating Room to Reason

In crafting a strategy to improve military decisionmaking, a logical starting point is to catalog cognitive abilities that could be useful in the intensity, complexity, danger, and urgency of war:

- perceiving reality objectively
- diagnosing complex problems
- knowing what information to seek
- differentiating between good information and bad information or noise
- absorbing and recalling critical information
- interpreting the behavior of others
- anticipating the behavior of others
- perceiving how the enemy senses and thinks
- forming a coherent view of unfamiliar and unfolding situations
- weighing diverse, competing, subjective values and interests
- setting achievable goals and metrics
- establishing priorities
- imagining feasible ways of accomplishing goals
- perceiving opportunities
- foreseeing the consequences of different courses of action
- analyzing the practicality, costs, and benefits of multiple options
- understanding and managing risks
- deciding provisionally despite information gaps
- admitting and remedying mistakes
- rethinking goals and adapting strategy.

Of course, what force commander would not want subordinate decisionmakers with such abilities? But the question immediately prompted by such a long list is how to identify those abilities that offer the greatest leverage in achieving military success in plausible operational situations. To this end, it is helpful to recall the military-operational conditions and challenges of the security environment shaped by the information revolution, the geopolitical revolution, and the access of adversaries to information-network technology. The salient features of this environment are:

- more and better information
- strategic and operational messiness
- networked, if asymmetrical, opposition forces.

In a landscape with such features, the value of being able to make good sense and use of information is self-evident. So, too, is the ability to use information to supplement mental models—in effect, to make complex reality more comprehensible when solving problems and making decisions. Yet while these abilities are worthy, they do not help much in pin-pointing priorities for improved cognitive effectiveness. For that, a deeper analysis of the dynamics of military operations is needed.

Again, blending intuition with reasoning is crucial in the environment described above: intuition, because of urgency; reasoning, because of change (unfamiliarity) and the high stakes of warfare. Enhancing time-information is crucial to creating room to reason, the better to solve complex problems and make difficult decisions urgently, which in turn can produce success in battles and wars. Thus, *the ability to use information to reason despite lack of time—to buy time and to make better use of time—can be of pivotal importance when time is scarce and information is bountiful.* Being able to turn an information advantage into a time advantage has even greater utility when growing complexity makes time more precious and reasoning more essential, as is true of the new security era.

This may be thought of as gaining a time-information edge over the opponent. Yet as adversaries incorporate IT and apply networking principles, gaining such an edge will not be easy. Those who understand the promise and principles of networking will seek their own time-information edge. Arguably, current operations in Iraq show that the terrorists and insurgents have a time-information edge, even though coalition and Iraqi government forces have better information-network technology, not to mention better firepower, at their disposal.[1]

It follows that special priority should be placed on those particular *cognitive abilities that can produce a time-information edge against networked opponents who are themselves aware of the value of such an edge.* If time can be turned consistently against enemy forces, an asymmetry in the opportunity to think can be achieved, which means an advantage in decisionmaking in the confusion and intensity of combat. In turn, superiority in decisionmaking can provide an edge in networked warfare. Therefore, those cognitive abilities that enhance time-information—that use information to turn time into an ally—deserve special attention.

This is not unlike the ways especially agile companies anticipate and affect the timing and speed of market changes, such as in introducing new products, putting competitors on the defensive, or forcing them to act in haste without adequate information, comprehension, or preparation. This ability is of greatest value in fast-moving, dynamic systems, such as IT markets. Those with cognitive time-information superiority can determine the rules of a constantly changing game.[2] While one associates such abilities with small and supple firms, bigger companies, such as Microsoft, also have exhibited such traits, dwarfing the competition. The challenge for the U.S. military is to"be like Microsoft"—big and powerful, but also a step ahead in understanding, shaping, and excelling in the cognitive realm of competition.

What are the cognitive abilities that matter most in networked warfare? One way to answer the question is to consider operations against networked forces. Apt examples can be found by looking again at China and al Qaeda.

Engaging Adversaries with Networked Forces

How could hostilities between U.S. and Chinese forces be affected as China employs IT, such as better, extended-range information, surveillance, reconnaissance (ISR) and networking concepts? While it is highly likely that U.S. forces will remain more advanced than Chinese forces for a decade or more the operational effects of networked forces that are up against other networked forces do not directly translate to force supremacy.

Networking provides three core benefits: enhanced *awareness*, thanks to the ability to acquire, fuse, and disseminate useful data from arrays of networked sensors; greater *precision* in weapons effects, thanks to target-location enabled GPS and other weapon-guidance information available via the network; and the ability to *integrate* dispersed units, platforms, and other force elements while managing them as a coherent force and concentrating their fires, thanks to broadband communications. Simply stated, networked forces are good at finding and destroying opposing, non-networked, forces while remaining safe from them.

But what about networked warfare, where the forces of both sides are networked? If two opposing forces possess all three networking benefits in some measure, the ability of each to operate safely while dispersed would be offset by the ability of the other to find and destroy its forces. Generally speaking, the ability to find and destroy forces (unless hidden) should prevail over the ability to make them survivable by dispersing them.[3] If this is so, the end result would be that both networked forces would be more vulnerable than if opposed by non-networked forces. The stronger, better networked force would, of course, be less vulnerable than and better able to find and destroy the inferior one. But that superior force would be more vulnerable and less effective than if it were opposed by a non-networked force, all else being equal.

In a hypothetical Sino-American conflict, networked and information-rich Chinese forces could well be less vulnerable to U.S. forces than they are now and also pose a greater threat. These are precisely the conditions the Chinese are striving to create. As they would be the first to admit, this would not mean parity in vulnerability, for Chinese forces would remain much less survivable than U.S. forces. But it would mean a convergence in vulnerability.

This is not some theoretical possibility of an imaginary future. As noted in the preceding chapter, China is now concentrating investment in the capabilities to increase the vulnerability of U.S. maritime power, which is the backbone of U.S. military power in the Western Pacific. While this would not necessarily make Chinese expansionism or Sino-American conflict likely, it could alter the power-political balance in the Western Pacific—a shift with potential strategic implications in one of the world's most vital regions. As China deploys and networks more and better sensors and precision weapons, and as it acquires the means and doctrines to operate dispersed forces, the efficacy of U.S. intervention against Chinese forces declines, while the risks rise.[4]

Under conditions of convergent vulnerability, each side would seek a decisive operational edge by tipping the force-vulnerability balance in its favor, even while trying to avoid dangerous escalation, particularly into the nuclear realm. As one RAND study notes:

> The more visible the battlefield, and the more that visibility is tantamount to destruction, the more difficult it will be to go to war with

platforms. . . . [Consequently,] exposure time must be short. . . . Ironically, a confrontation between two advanced, net-centric militaries will likely reduce the importance of technology in favor of people and their ability to make rapid but accurate decisions with incomplete or overwhelming amounts of information.[5]

Time-information superiority is crucial to victory in networked warfare, and the key to gaining it lies in battle-wise thinking and decisionmaking. "Victory, if not inherent in the balance of forces or unique attributes of geography, falls to whoever has the best combination of surprise, error control, fortune, and highly trained people."[6] In the case of any future Sino-American conflict, the geographic disadvantage of U.S. forces could be aggravated by a loss of U.S. networking supremacy.

Being increasingly vulnerable to networked Chinese sensors and precision weapons, U.S. decisionmakers would need to startle and stun Chinese counterparts by catching them flat-footed, striking from surprising directions and distances, maneuvering in surprising ways, reacting quickly, doing the unexpected, shunning the expected, and expecting the unexpected. They would need to anticipate Chinese moves intended to increase or exploit the vulnerability of U.S. forces, lure Chinese forces into actions and engagements that increase their vulnerability, operate in changing configurations, and display uncommon tactical flexibility—all by virtue of superior networked decisionmaking. U.S. decisionmakers would need to be several moves ahead of Chinese decisionmakers. To maximize sudden opportunities, U.S. commanders would have to encourage local initiatives and enable deployed field grade officers to make decisions on the spot. The resulting operational advantages could be significant, particularly if Chinese officers must call Beijing for instructions. Throughout any operation, U.S. forces and decisionmakers would seek to get time on their side—as would the Chinese, presumably.

Unlike the case of gradually convergent Sino-American military vulnerability, the main challenge in direct military or other operations against al Qaeda is to increase the terrorists' current relatively low vulnerability and the advantages that go with it. The time between spotting and raiding a terrorist cell must be shrunk to less than the time it takes the terrorists to flee once they become suspicious. Delay is more likely to be the result of human indecision or inability to make sense of information than of slow physical movement of forces. Cognition and decisionmaking can be sped up to some extent by standardizing operating procedures and tactics.

On the other hand, predictability can be as fatal as tardiness in combating terrorists.

Al Qaeda's forces are arrayed asymmetrically against its opponents. Besides its gross inferiority in firepower, al Qaeda does not even possess its own IT, relying instead on generally available sources of information and communications infrastructure. It cannot begin to challenge the information superiority of U.S. and other Western security forces. Yet it is capable of *time-superiority* advantages, from its ability to wait and plan patiently before striking to its agents' ability to slip away quickly with little warning from forces seeking to subdue them. Whether deliberate, opportunistic, or quick, al Qaeda's ability to use time as an ally is inherent in its modus operandi. Moreover, its decisionmaking may be simpler than that of its enemies. All things considered, al Qaeda may now hold a time-information edge, which reduces its vulnerability compared to that of its targets.

With the cases of China and al Qaeda in mind, what may happen is illustrated, in a purely notional fashion, in figure 4–1. All else being equal, forces that are networked can dominate those that are not, as shown in lower-right and upper-left corners. For example, the highly networked U.S. forces in Afghanistan and Iraq were initially able to dominate their minimally networked opponents (A). However, as opposing forces become *more* networked, the invulnerability and thus operational dominance of U.S. forces could erode (B). This would create a convergence of enemy and U.S. vulnerability, even though U.S. forces would remain stronger and superior in their use of networking. Both China and al Qaeda could minimize U.S. dominance and increase U.S. vulnerability by networking, leaving U.S. forces (and security) worse off in the future than they are at present.

Despite the differences between the al Qaeda and China cases, the operational centrality and sensitivity of time, timing, timeliness, and time-information stand out. The cognitive abilities to act first, react quickly, choose the right time to act, catch the enemy off guard, learn and adjust quickly, and deprive the enemy of time to learn and adjust will be increasingly valuable as vulnerabilities converge. The exploitation and manipulation of time can tip the vulnerability balance significantly, if momentarily. With effective cognition, networked information can be used to find and use time.

Figure 4–1. The Shift toward Convergent Vulnerability

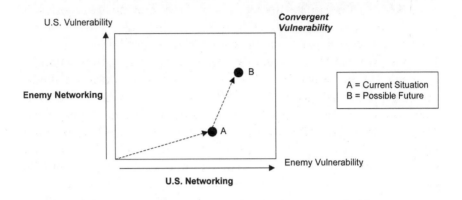

Some of the same time-sensitive abilities that matter in the China case are important in military or law-enforcement operations against terrorist networks: anticipating the enemy's next tactic, weapon, and target; surprising the opposition; doing and expecting the unexpected; executing swift decision and reaction times; and missing no opportunity. In all cases, the side that is superior at translating information into time-information advantage can gain critical operational advantages under conditions of convergent vulnerability. This is the purpose and payoff of battle-wisdom.

Being battle-wise means using information to master the urgency of war. By expanding and exploiting the opportunity to reason, battle-wisdom can turn time into the enemy of one's enemy. In the case of China, it could mean tipping the balance of vulnerability between Chinese and U.S. forces by leaving Chinese forces exposed or paralyzed at critical moments. In the case of al Qaeda, it could mean shortening, just enough, the time lapsed between receiving word of the terrorists' whereabouts and launching a strike or raid against them.

Critical Cognitive Abilities in Networked Warfare

This glance at operations in conditions of convergent vulnerability allows us to spotlight four critical abilities that could separate winners from losers in networked warfare: *anticipation, decision speed, opportunism,* and *rapid adaptability.* All four optimize cognition and time:

- Anticipation can provide initial advantage and set conditions for success.
- Speed in making and executing decisions can exploit these conditions.
- Opportunism can yield sudden, nonlinear advances and advantages.
- Rapid—or real-time—adaptability can improve performance in the light of information gleaned from the current battle.

These particular abilities depend on the synergy of reliable intuition and timely reasoning and may be employed to increase the exposure time of enemy forces and reduce that of one's own—a key factor in tipping the balance of vulnerability to one's advantage. Each involves using information to gain time-information advantages. Together, they can provide a mental and operational edge from the commencement to the conclusion of hostilities. They are closely akin to the abilities of the superlative basketball player whose "court sense" and speed of mind and foot leave the opposing team playing in what looks like slow motion while elevating his own team. Such cognitive abilities provide an edge when physical attributes may not. They define the type of superior decisionmaking that yields superior performance during intense and urgent conflict against a capable opponent.

A force with people—individuals and teams—who have these key abilities and are able to use them to gain time-information advantage can cause less battle-wise opposing forces to seem, in effect, more "mechanized." For U.S. forces, superiority in these cognitive abilities would have the effect of depriving adversaries of the benefits of networking (see figure 4–1). Viewed strategically, the loss of a monopoly in military networking would be offset by the ability of U.S. forces to attain time-information primacy with these key cognitive abilities—thus getting the most out of their own networking and negating that of enemy forces.

Yet the China and al Qaeda cases suggest that such cognitive abilities may be possessed by both the United States and its adversaries. While the United States is fortunate to have many military advantages—not only in technology but also in the general intelligence, education, training, and discipline of the men and women in its armed forces—it is unclear that its soldiers have unassailable superiority in their anticipation, decision speed, opportunism, and rapid adaptability. Therefore, the strategic goal should be not just improved battle-wisdom but superior battle-wisdom.

To illustrate the importance of these cognitive abilities, imagine SOF inserted into a remote ungovernable region of Central Asia that has become a terrorist haven. The SOF are networked with precision sensor- and weapon-bearing drone aircraft and other intelligence and strike capabilities, and they have hired local scouts and informers. They are seasoned, disciplined, and skilled fighters, and they can move quickly. The terrorists are poor in technical sensors but have better human sensors than the SOF. They are dispersed yet able to communicate with one another, though at some risk of discovery. Learning they are up against SOF, the terrorists' aims are to survive and kill another day, not to stand and fight.

The SOF, in this case, rely heavily on experience, for example Afghanistan, in which the terrorists fought first and then fled. Anticipating this same pattern, SOF commanders concentrate on preparing for combat and closing with the enemy. Consequently, they miss a fleeting opportunity to cut off the terrorists' escape routes. Instead of adapting swiftly upon the first hint that reality is at odds with their intuition, SOF commanders deliberate while waiting for conclusive data from their remote sensors. The time it takes them to gather and analyze information and then adjust their plan is just enough for the terrorists to melt into the wilderness and tribal populations, to wait and plan new terror. Because of cognitive failure, superior and better networked SOF lose the time-information advantage and, therefore, the bloodless battle.

A complete definition of battle-wisdom thus can now be considered: the melding of reliable intuition and efficient reasoning to improve anticipation, fast reaction, opportunism, and quick adaptability for time-information superiority in complex, intense, and possibly confusing networked warfare—more simply, creating time to think and decide wisely in the midst of violence.

Operationally, the aim of developing battle-wise forces is to foster these four key abilities as a way of gaining and holding a warfighting edge even in conditions of convergent vulnerability. Strategically, it is to ensure that the United States can use force as its security interests and responsibilities warrant, despite the loss of its monopoly in networked warfare. While having battle-wise forces and people does not guarantee the achievement of these objectives—many other factors are involved in warfare in this new era—it can improve the odds.

To some extent, such battle-wise abilities were in evidence during the opening intense combat phase of the conflict in Iraq. The opening moves of U.S. and coalition forces involved operational initiative and surprise, even though no one doubted an invasion was imminent. Still the forces came from many directions, not only the expected one. When splinters of Iraqi forces waged rear-guard actions, coalition forces quickly adapted by carefully dividing attention between extinguishing this unexpected threat and taking the capital. Coalition forces approached Baghdad cautiously, but then acted quickly when a foray into the city revealed the opportunity for a coup de grace. The only ability not evident was anticipation, in that U.S. and coalition forces expected a "regular" defense by Iraqi divisions but encountered a mix of regular and irregular forces and tactics.

After the initial phase, however, irregular insurgent forces, joined by foreign terrorists, scattered and hid among the population to minimize their own vulnerability and ambush U.S. forces. The foreigners, while few in number, are key, for they add the cognitive talents of al Qaeda to a collection of Saddamist killers with no experience in networking and no record of mental acuity or agility. Against this mutating threat, the same U.S. forces generally have not reacted with the anticipation, decision speed, opportunism, and rapid adaptability they showed earlier against much larger but hierarchical Iraqi forces. Yet these are precisely the battle-wise abilities U.S. forces need to prevail against such threats.

Can there be any doubt which side showed better cognitive skills and had a time-information advantage in the first phase—or the second? This suggests several points. First, U.S. forces can exploit networking much better when up against non-networked forces than networked ones. Second, even the remnants of the slow and bloated armies of Saddam Hussein can gain advantages by dispersing and networking. Third, the introduction of elements with strong cognitive abilities—foreign terrorists, in this case—created a more formidable opponent for U.S. forces. Fourth, cognitive superiority is difficult but, if anything, more critical when facing networked adversaries, however inferior they may be by traditional measures.

Complexity and Simplicity

The value of being able to use information to gain time at the adversary's expense is related to another factor in this analysis: complexity. As Herbert Simon observed, human problem-solving suffers when the complexity of causes and effects of real-world dynamic systems increases. Because reasoning takes time, the greater the complexity of a situation, the more precious time becomes and the shorter it may seem—in effect, the shorter it is. Therefore, as complexity increases, it becomes more essential to use information to gain time and to use that time economically to reason. It is hard to think of a human activity in which this is more relevant than warfare.

Battle-wise warfighters not only can handle this nexus of complexity, time, and reasoning, but they can also exploit it. Confronting enemy decisionmakers with added complexity can, in effect, starve them of time, undermine their reasoning, and degrade their decisionmaking. The abilities to anticipate, decide quickly, see and seize opportunities, and adapt rapidly can have this effect. Because cognitive effectiveness is of such importance in networked warfare, battles could be won by complicating the adversary's reality. Of course, increasing the complexity facing the adversary has even greater leverage if done while reducing the complexity facing oneself.

Once more, the China case provides a fitting example. Quick and unexpected actions, anticipation of or prompt response to Chinese actions, or shifts in the tactics of U.S. forces, could overload the ability of the Chinese to comprehend what is happening and thus make sound decisions. This could shrink Chinese time-information, expand U.S. time-information, and help tip the balance of vulnerability more sharply in favor of American forces.

Such measures might be considered psychological operations (PSYOP) or information operations (IO). It is no coincidence that PSYOP in particular and IO in general are becoming increasingly popular as information and cognition figure more importantly in warfare. However, whereas PSYOP and IO are auxiliary tools used to complement the operations of forces, battle-wisdom governs how the forces themselves are used. It would be one thing to confuse Chinese commanders by infecting their computers and quite another to confuse them by suddenly changing the nature of strike operations or catching their forces at their most vulnerable. In any case, as Chinese forces incorporate IT and networking, battle-wisdom, PSYOP, and IO all will become increasingly critical for U.S. forces.

Al Qaeda is already acting in ways intended to complicate U.S. decisionmaking and, in effect, compound the time-information problem facing American military and other counterterrorism forces. The ingenuity of the 9/11 attacks, and the fact that U.S. decisionmakers were baffled by such a complex, unfamiliar, and urgent threat, show the potential leverage to be gained by using unconventional means, in this case commercial jets, as weapons. The terrorists' use of surprise not only as to the timing but also as to the method of strike yielded a decisive time-information edge. And, of course, no experience or mental model was of much use to U.S. leaders. Since 9/11 and the ensuing U.S. action in Afghanistan, al Qaeda has augmented its cognitive arsenal with the use of false information designed to overwhelm U.S. counterterrorism efforts.[7]

Gaining a time-information advantage over al Qaeda will not be easy, given its distributed form, skill at concealment, and utter lack of inhibition in regard to targets and weapons. The question is whether and how the United States and its antiterrorist allies can add to the operational complexity faced by al Qaeda and its agents.

At this point it may be useful to return to the case of the SOF unit that failed to capture the terrorists in Central Asia—this time changing the thinking of the SOF. What if the SOF distrusted the relevance of their experience and the reliability of their intuition enough to be alert to any indication that these particular terrorists might choose to flee instead of fight? By anticipating and adapting without delay, the SOF may move to cut off every escape route—which SOF can do in very small, quick, networked units. At this point, the problem facing the terrorists can become mind-boggling, if not hopeless. Should they attempt to escape by attacking smaller SOF units, fight in place, split up, hide, surrender, seek more information—or perhaps just keep thinking and waste critical time? Moreover, having bought time, the SOF could gather more information from sensors and scouts to learn what the terrorists were plotting and then analyze whether and how to engage or ensnare them. This information would, in turn, allow the SOF to control the timing and conditions of engagement. The SOF could think and act based on a clearer view of reality, while the terrorists could be confounded by conflicting reports from their agents about SOF actions. Hence complexity would increase for the terrorists and decrease for the SOF.

Of course, the technical ability of the SOF to network with each other, with sensors, and with back-up strike forces is what makes such an operation possible. Why not make the most of it by gaining an edge at the cognitive level as well? Networking is necessary but not sufficient for the SOF in this illustration. The deciding factor is battle-wise exploitation of information.

Battle-wisdom goes deeper than reliable intuition, timely reasoning, and the cognitive abilities suggested here. Character matters. Soldiers are more likely to be battle-wise if they are willing to learn and take responsibility for the effects of their decisions. Recognizing these prized traits—to learn and to lead—is crucial in finding, developing, and using battle-wise decisionmakers, regardless of rank. Just as lieutenants must be willing and able to lead, lieutenant generals must be willing and able to learn.

The Race Is On

In the age of networked warfare, learning in action is as important strategically as it is operationally and tactically. The long lag-times before institutions take full advantage of information networking are well known. This is so in part because overhauling operational processes and organizational structures takes time. In contrast, people can adapt to new information devices and programs very quickly. Yet people within institutions, such as the armed services, need more time to change the way they think and act after having mastered the technology itself. We are accustomed to steep curves of improvement over time in the performance of IT, but the curve that describes the rate at which the technology yields higher levels of human performance can be much flatter.[8] The implication of this is that the lead of U.S. forces in incorporating IT and adopting networking principles may not translate into a substantial and lasting operational lead over adversaries that also embrace networking.

In its own way, and relative to its own purposes, al Qaeda may already be ahead of the security forces the democracies have arrayed against it. It has patterned itself according to networking principles, is comparatively invulnerable, is working intensely on its own version of battle-wisdom, and mutates as needed. While it has not recently mounted an operation as spectacular as 9/11, al Qaeda has created havoc in Iraq, attacked power centers of the Saudi monarchy, spawned affiliates, inspired copy cats, and branded itself the model of new terrorism. Even while commandeering global information technology, infrastructure, and mass distribution for its purposes, al Qaeda appears to understand, at least implicitly, that how its agents think—fanatical yet disciplined, rational yet intuitive—is key.

Figure 4–2. Battle-Wise Learning Curves

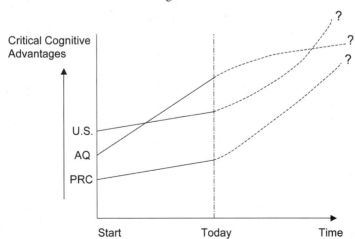

Figure 4–2 depicts notionally the rate at which U.S. forces, Chinese forces, and al Qaeda may be exploiting the potential of networking to gain operational advantages relative to their own ends and strategies. (No direct comparisons are implied—or, for that matter, really meaningful.) Not long ago, al Qaeda faced a steeper curve than U.S. forces in learning how to exploit networking to its ends, but it may now be—it is of course hard to know—climbing its curve faster than the U.S. military establishment is climbing its own. In its own way, al Qaeda has gained ground by exploiting its particular form of networking with its type of battle-wise people. China, while moving slowly, could begin accelerating any time.

Of course, there is no way of accurately measuring the distance between or the slope of these curves today, much less projecting them into the future. We suggest them only to provoke thought and bring attention to the potential for adversaries to exploit IT and networks for cognitive gains. One important message is that the rate of learning how to exploit the technologies is more or less independent of the ability to invent them and the resources to build them. Another is that gains in the abilities of adversaries to think, decide, and act—thanks to networking principles and capabilities—must be viewed not only in relation to U.S. abilities to do so but also to the purposes and strategies of those adversaries.

This chapter will close with two chilling quotations—one about al Qaeda by a renowned expert on security in the information age, the other about China from the most recent official statement of Chinese military strategy:

> While al Qaeda may look amorphous (i.e., shapeless), the deeper reality may be that it is . . . deliberately shifting its shape and style to suit changing circumstances, including the addition of new semi-autonomous affiliates to the broader network. . . . Al Qaeda is using the information age to revitalize and project ancient patterns of tribalism on a global scale.[9]

> Based on the transformation of modern warfare . . . the PLA develops its military theories in an innovative spirit and explores . . . conducting operations under the condition of informationalization. In accordance with the principle of making the troops smaller and better, as well as more integrated and efficient, and with emphasis on adjusting organizational structure and reforming the command system, the

PLA works to build and further improve the military structure and organization to make them ... flexible and swift in command.[10]

It would be a capital mistake for the United States to assume that an insurmountable lead in the theory, science, technology, production, and use of IT and networking guarantees a lead in battle-wisdom and the operational advantages that come with it.

Notes

[1] The main reason U.S. troops in Iraq suffer from a time-information disadvantage is the difficulty in getting information about the insurgents from the Iraqi people.

[2] We note the work of Stanford University economist Brian Arthur in understanding and explaining how cognition of dynamic market-competitive conditions can be and has been exploited, especially in the fast-moving IT markets.

[3] Of course, it is possible to think of many exceptions to this generalization, for example, by shifting maritime fire-power to subsurface vessels, by moving dispersed forces at speeds high enough to frustrate detection and destruction, and by multiplying possible weapons platforms to present too many targets.

[4] The fielding of long-range surveillance capabilities may be the "long pole" for the Chinese because without these capabilities, their forces, however well networked, will be blind to U.S. forces beyond coastal waters. The Chinese know this and are investing in space-based and other sensors.

[5] James C. Mulvenon et al., *Chinese Response to U.S. Military Transformation and Implications for the Department of Defense* (Santa Monica, CA: RAND, 2006).

[6] Ibid.

[7] George Friedman, *America's Secret War: Inside the Hidden Worldwide Struggle between America and its Enemies* (New York: Doubleday, 2004).

[8] Compare the 12–18 months it takes for microprocessor performance to improve by 100 percent to the decade or more before investments in this technology produced noticeable increases in economic performance and productivity.

[9] David Ronfeldt, "Al Qaeda and Its Affiliates: A Global Tribe Waging Segmental Warfare?" *First Monday* 10, no. 3 (March 2005), available at <www.firstmonday.org/issues/issue10_3/ronfledt/index.html>.

[10] State Council Information Office, *China's National Defense in 2004* (Beijing: State Council Information Office Press, 2004).

Chapter Five

Integrating Intuition and Reasoning in Action

A thread that connects key cognitive abilities is the capacity to learn in action. Because of its intensity and tempo, warfare generates information rapidly, and networking accelerates it. Combined with the messiness and ambiguity of combat, this rapid flow of information can cause confusion or, in military vernacular, fog. At the same time, being able to think, decide, and act quickly has obvious advantages. Again, time-information shows its significance. Deferring decisions until all useful information is available and analyzed can make a force too slow. Yet making irretrievable judgments with deficient information can lead to casualties and calamities. Decisionmaking depends heavily on the balancing and management of time and information. Battle-wisdom must apply not only to people but also to how those people make decisions.

In networked warfare, decisionmaking should be based on:

- knowing what can and must be decided, and when
- making provisional decisions pending more information
- using provisional decisions to gain time and information
- revisiting decisions as more information is harvested.

Such an approach can be at once expeditious and thoughtful. Done right, it can master the urgency of war without compromising performance by either haste or delay. It can expand the room for reasoning—relying mainly on intuition when a challenge suddenly arises, but then shifting toward reasoning as time and information are gained by provisional decisions and actions. With battle-wise decisionmaking, the warfighter times and tailors choices to take account of the need and opportunity to learn in action. While learning occurs, the warfighter can confirm, improve, alter, or even reverse provisional decisions. Just as the ability to adapt rapidly is a key strength of the battle-wise warfighter, the practice of learning in action is a key feature of battle-wise decisionmaking.

While learning under pressure may sound straightforward, learning while fighting is exceedingly hard and potentially hazardous. Warfare and reflection are uneasy companions. Yet battle has no substitute when it comes to the opportunity to learn about both the opposing force and one's own force. As coaches and players do during an American football game—also dynamic, messy, violent, and networked—the battle-wise decisionmaker starts with a flexible plan based on prior experience and analysis and, as the situation unfolds, decides what "plays" to execute based on what is learned from the action. Neither a rigid plan nor pure intuition will do. Just as battle plans may not survive beyond initial contact with the enemy, learning lessons need not await the end of hostilities.

In recent years, researchers at the Santa Fe Institute, the RAND Corporation, and elsewhere have formulated important precepts and championed useful planning methods based on the belief that complexity and uncertainty are best addressed by adaptive strategies for the long-term future.[1] The underlying idea is to understand what one *must and can* decide, depending on urgency and available information, while playing for time and seeking more information to improve the quality of decisions. The same approach, drastically compressed in time, should underpin battle-wise decisionmaking. Mastering it could provide stunning operational and enduring strategic advantages.

Again, both intuition and reasoning are indispensable in overcoming the pressure, urgency, and messiness of warfare in the information age. Sensing a threat or an opportunity, initial action may be based on what experience says ought to work, but should also aim to gain both information and time. As this action clarifies conditions and buys time, structured reasoning becomes more possible, leading to a refined or revised course of action after examination of options. Along the way, the decisionmaker looks for signs that should appear if preceding assumptions and judgments were correct. Not seeing these signs may signal that a course correction is needed. Meanwhile, information can be used to refine understanding and adjust accordingly. Eventually, with time and information now on the decisionmaker's side, sound and superior reasoning can lead to success.

This decisionmaking process is depicted in figure 5–1, which illustrates the increase in confidence with the passage of time. The four distinct conditions shown are somewhat artificial; in reality, the process is more continuous and fluid. At any point in the process battle-wise decisionmaking should offer major advantages, all else being equal, over opposing forces that are guided by decisions dictated by either haste or undue caution.

Figure 5–1. Battle-Wise Decisionmaking Process

Condition 1	Condition 2	Condition 3	Condition 4
Limited grasp of the problem	Clarification of the problem	Comprehension of the problem	Complex solution
Urgent	Less time pressure	Little time pressure	No time pressure
Little information	Limited information	Abundant information	Complete information
Act to create options	Act to gain advantage	Act to gain dominance	Succeed
Move in general direction	Refine or alter direction	See desired end state	End state
Intuition-heavy	Intuition & Reasoning-lite	Reasoning & Intuition-lite	Reasoning

- **Gain information**
- **Reduce time pressure**
- **Refine objective and course of action**
- **Increase the role of reasoning**

One of the keys to integrating intuition with reasoning is the *self-awareness* of the decisionmaker.[2] Knowledge of the origins, assumptions, biases, and limits of one's mental models and experiences can help answer the question: Does my intuition apply to the situation I face? If the answer is yes, intuition may be a reliable basis for deciding at least provisionally. But if there is no suitable mental model or body of relevant experience, more information should be sought and analyzed before making a decision. Even then, intuition may be helpful to borrow some time.

Because people who must act with little time naturally favor intuition over reasoning, a conscious intervention—verbalizing the applicability question—may be needed to avert mistakes when intuition is inadequate or misleading. The individual must be able to determine dispassionately whether stored models apply. Such disciplined and objective self-awareness is hard for most people, but it can be cultivated.

Battle-Wisdom in Practice

In warfare, the ability to make sense of information is already critical and will become more so.[3] Networking provides rich but also potentially confusing information. To some extent, the richness can be exploited and the confusion reduced by technologies that sort, distribute, and display data. Although this may serve up better information, it does not assure its effective use. This is why one of the prerequisites of battle-wise decision-making is the practical implementation of smart pull.

In practice, it is far more difficult to satisfy via the smart-pull method the information needs of a warfighter than those of the average Internet user. In the first place, knowing what information to post on a military network implies knowing all facets of all predicaments and opportunities a warfighter—for that matter, *all* warfighters—may face and thus what information might be helpful. Even then, the warfighter will not know of all the relevant information that could be pulled. The image of the unit commander under surprise attack having to browse the operations network for useful data, as one would look for cheap flights or long-lost classmates, is hardly comforting. So the expectation that a local decisionmaker will, in fact, have all useful information requires a great deal of faith, or else investment in network development aimed at providing decisionmakers with useful information in a usable form at the right time.

Although network development to this end is beyond this volume's scope, it is obviously important and, for military networking, hardly straightforward. Even smart users trying to pull information from networks are hindered by chaotic, messy conditions and time constraints. To be effective in operations, the design and operation of networks must take account of the predilections, culture, habits, nomenclature, and contingent needs of myriad users under myriad conditions. Even then, the value of data can be undermined without the situational context of a problem, which in military operations may be unforeseeable. Finally, the requirements and efficacy of network capabilities vary with the decisionmaker's experience. Experienced individuals know what information to select, have a more coherent mental organization of information, recognize what information is missing, and are able to adjust decisions to compensate for incomplete information. Yet no condition could be less amenable to the orderly use of networking than combat. A warfighter engaged with an enemy is not, and probably could never be regarded as, the equivalent of an ordinary Internet user.

Under current conditions, senior commanders can have remarkably comprehensive information displayed in exquisite detail before them. But whether they have the necessary confidence in subordinates to delegate authority and the self-discipline to resist micromanagement, despite these seductive displays, remain important open questions. Of course, confidence must be earned. The subordinate must be not only battle-wise but also willing to take responsibility for the consequences of his or her decisions—to be a leader-in-action, if not in pecking order.

Until enough tactical-level officers are sufficiently battle-wise to make good use of the information from the network, senior commanders will understandably be reluctant to delegate and tempted to micromanage. For tough decisions, a good leader would rather risk making a mistake and taking the fall than having a subordinate do so. While other military cultures have long stressed decentralized operational decisionmaking, U.S. senior officers likely will set the bar high for battle-wise juniors to earn such authority.

In theory, battle-wisdom demands the integration of intuition and reasoning, self-awareness, the abilities to anticipate, decide quickly, seize opportunities, and adapt in action, and the willingness to lead and learn. In practice, it also depends on implementation of the smart-pull principle, good information management (IM), and delegation of authority.

An Illustration

An example may illustrate these factors at work: imagine that a motorized column of peacekeepers is ambushed as it moves through a remote province of an African country engulfed in tribal violence. Imagine that this unit is networked with nearby patrols, sensor-carrying drone aircraft, an attack-helicopter unit, a provincial operations coordination center, force headquarters, and an intelligence fusion facility. Now visualize the major in command of the ambushed column being not at the network's edge but at its center. Assume that good IM is in place and that this battle-wise officer is trained to know what information to pull from the network, including intelligence about the threat, the latest data on the noncombatants to be rescued, weather reports, and information about the availability of backup forces.

Senior officers up the chain of command feel the major has earned their confidence, and they appreciate that he has a fuller immediate view than they, thanks to his tacit knowledge, of unfolding events. Therefore, within the unit's stated mission and rules of engagement, the major has the authority to decide how to respond and what support to request. If it were possible that a wrong decision by the major could jeopardize not only his own unit and mission but also the larger operation or other units, an overriding decision by higher command might be indicated.[4] This often will be the case, given the ease and speed with which word of the unit's fate is shared with the outside world. But in this example, the assumption will be that the major's chosen course of action, whether right or wrong, will not have major ramifications beyond his unit and its results. He is thus inside his envelope of discretion.

Depending on what the major senses and summons from experience, his initial choice may be simply to hunker down or pull back. His intuition may tell him that the option of attacking the ambushing forces is a poor one because his experience and mental model suggest that an inferior force would not have attacked him. Once he has more data via his network about the threat, the presence of innocent civilians, and the time it will take to be reinforced, the major can weigh and decide among several options: to engage in a firefight; wait for reinforcement before engaging; retreat; or attempt to slip the ambush and proceed with the original assignment. While he may get advice from headquarters, the major is best placed to determine what is happening, what information and help is needed, what options are available, and what risks exist. The critical question then becomes how, and how well, he selects the best course of action. While vital, intuition will get the officer only so far before he must analyze all available information and weigh his options.

The major's self-awareness establishes that his intuition is reliable enough to tell him not to attack, but only that. Identifying, weighing, and selecting among options beyond "don't attack" require more information and more time, which he gains by holding his ground. Thus, his intuition can be trusted to give a good-enough initial response as well as secure him time to pull information from the network to aid in making a more reasoned decision.

The illustration shows, again, that the most precious commodities in situations of urgency and complexity, like warfare, are time and information. Yet the case reminds us that time and information often work against one another: The greater the haste, the less chance one has to process data and to reason, thus forfeiting the benefit of information technology. Lack of time means lack of information, and lack of information means dependence on experience and mental models, which may not be appropriate or sufficient to the unfamiliar problem at hand. However tolerable in routine problem-solving, this time-information problem must, and increasingly can, be overcome when lives depend on solving complex and unfamiliar problems.

Summing Up Battle-Wisdom

Now that we have examined the need for battle-wise people and decisionmaking, in strategic as well as operational terms, and before looking at how to foster these abilities, it is useful to tie together the concepts at play. Figure 5–2 presents a schematic of battle-wisdom at work, from the operational conditions that demand it, the abilities it comprises, the traits and conditions that foster it, and its payoff in networked warfare.

Figure 5–2. The Battle-Wise Process

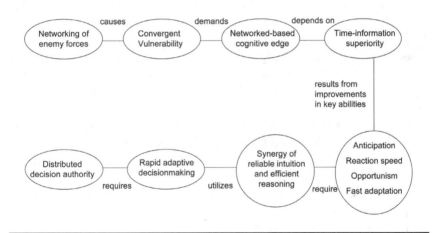

Note that once enemy forces are able to achieve networking and make U.S. forces vulnerable, U.S. forces are not guaranteed advantages in cognitive qualities, the ability to gain time-information advantage, the specific capabilities that may determine operational success or failure, the blending of intuition and reasoning, rapid adaptive decisionmaking, or the distribution of authority to go with the distribution of information. More generally stated, *once two competitors have both crossed a certain threshold of networking, their comparative ability to achieve better cognition and decisionmaking is not determined by their economic, technological, or military resources.* Shrewd, adaptable, and strongly motivated adversaries, whether technologically sophisticated or not, could be competitive with the United States in each of these elements.

This implies that *the United States must fashion and follow a strategy that does not presuppose battle-wisdom superiority.* To mix athletic metaphors, cognitive competition in the military domain is a new ball game on a level playing field. Therefore, the U.S. strategy must not be a mere extension of efforts already under way to create network-centric forces but instead an explicit and comprehensive plan to compete and win at the next, higher level.

Has the U.S. Military Grasped the Need for a Strategy?

Sound thinking and decisionmaking are more than mere loose ends of network-centric warfare. Moreover, success in competition on the military cognitive plane will not necessarily follow success on the technological plane. Therefore, what is needed is a coherent strategy to build battle-wise forces. Do the makings of such a strategy now exist?

Readings of the latest *National Military Strategy of the United States of America* suggest that the answer is a qualified no. Here are germane excerpts from that document:

> Decision superiority—"the process of making decisions better and faster than an Adversary"—is essential to executing a strategy based on speed and flexibility. Decision superiority requires new ways of thinking about acquiring, integrating, using and sharing information. It necessitates new ideas for developing architectures for command, control, communications and computers (C4) as well as the intelligence, surveillance and reconnaissance assets that provide knowledge of adversaries. Decision superiority requires precise information of enemy and friendly dispositions, capabilities, and activities, as well as other data relevant to successful campaigns. Battlespace awareness, combined with responsive command and control systems, supports dynamic decisionmaking and turns information superiority into a competitive advantage adversaries cannot match.

> Persistent surveillance, ISR management, collaborative analysis and on-demand dissemination facilitate battlespace awareness. Developing the intelligence products to support this level of awareness

requires collection systems and assured access to air, land, sea and space-based sensors.

Decisions to apply force in multiple, widely dispersed locations require highly flexible and adaptive joint command and control processes. Commanders must communicate decisions to subordinates, rapidly develop alternative courses of action, generate required effects, assess results and conduct appropriate follow-on operations.

A decision superior joint force must employ decisionmaking processes that allow commanders to attack time-sensitive and time-critical targets. Dynamic decisionmaking brings together organizations, planning processes, technical systems and commensurate authorities that support informed decisions. Such decisions require networked command and control capabilities and a tailored common operating picture of the battlespace.[5]

These passages tell us there is a general awareness of the growing operational and strategic importance of something called "decision superiority," which has some but not all of the elements of the superiority in cognitive capacity and performance that we call battle-wisdom. It also tells us that responsive command and control systems, collaborative analysis, and on-demand dissemination of information are important to decision superiority. This is encouraging.

However, this official explanation of what is required for decision superiority fails to stress human cognition—how people think and how well they decide. It is as if battle-wisdom—the capacity to integrate reliable intuition and rapid reasoning and the abilities to anticipate, decide quickly, seize opportunities, and learn in action—is assumed, needing only better intelligence sensors, information networks, and processes to succeed. It calls for commanders to communicate their decisions to subordinates, without recognizing that the subordinates may well be better informed than their superiors to decide what to do. After all, the great virtue of networking is not that it enables commanders to promulgate orders but that it informs those "on the edge" and permits them to collaborate, accept responsibility, and take initiative.

The key to decision superiority lies not in the information network behind the screen but in the human brain behind the eyes looking at the screen. If the United States expects to lead on the cognitive plane the way it has led on the technological plane, it would do well to begin with a basic understanding of the difference between the two.

Notes

¹ We draw loosely from a growing body of work on adaptive planning. See Robert J. Lempert, Steven Popper, and Steven C. Bankes, *Shaping the Next One Hundred Years: New Methods for Quantitative, Long-Term Policy Analysis* (Santa Monica, CA: RAND, 2003). Whereas their work deals with long-term planning and decisionmaking, we advance the proposition that this way of solving problems may be compressed into operational time-frames.

² *Self-awareness* is defined as consciousness of one's own individuality, including the strengths, weaknesses, range, and limits of one's cognitive abilities.

³ Dennis K. Leedom, *Sensemaking Symposium Report*, with the Command and Control Research Program, Vienna, VA, 2001, available at <www.dodccrp.org/events/2001/sensemaking_symposium/docs/FinalReport/Sensemaking_Final_Report.pdf>.

⁴ In traditional terms, it may be helpful to think of the major's decision domain as being at the tactical level, whereas his superiors are responsible for the operational and strategic levels. While the value of these distinctions is being eroded by complexity and networking, they still adequately connote the levels at which decisions should be made.

⁵ Joint Chiefs of Staff, *The National Military Strategy of the United States of America: A Strategy for Today; A Vision for Tomorrow* (Washington, DC: Department of Defense, 2004).

Chapter Six

From Networking Power to Cognitive Power

T he spread of IT and networking concepts and the changing nature of warfare argue for giving more attention and resources to the improvement of human problem-solving. The significance of the human in networked warfare can be seen in two contrasting ways. The first is to regard people as the *weak nodes* of any networked system: unable to cope when deluged with information; notoriously bad at solving complex problems rationally—an impediment to technology fulfilling its potential. The other perspective is to view people as the *strong nodes* of a networked system; the network merely absorbs, processes, and moves data, leaving people to do what information systems cannot—make hard and responsible choices. We obviously subscribe to the latter view. But either way, the capability to distribute information has brought the question of cognitive power to the fore. Improving decisionmaking—creating battle-wise superiority—deserves attention not as a peripheral detail or afterthought of networked warfare but as its ultimate differentiator.

Improving cognitive effectiveness for networked warfare is not a sufficiently high U.S. defense priority. Apart from investment in information networks themselves, most effort is being directed at technologies and techniques to manipulate information to make it more useful to humans. Managing military information—collecting, fusing, filtering, processing, packaging, sharing, and displaying it—is important and challenging and merits the heightened attention it is getting. However, managing information is not the same as strengthening cognition; it happens outside the brain, not within it. Information management can help minds work but cannot do the work that minds must do and are best equipped to do. Unless this distinction is clearly made, we will confuse work on better displays, video-teleconferencing, and chatrooms with a strategy to improve thinking and decisionmaking.

What, then, are the elements of a strategy to gain cognitive advantage in networked warfare? As a starting point, a network can be thought of in three ways:

■ as a *distributor* of information to individual minds
■ as a *mobilizer* of many minds
■ as a *venue* for collective thinking.

Accordingly, there are three ways to enhance cognitive contributions to military success:

■ improving the ability of individuals to make use of distributed information in thinking and deciding
■ empowering more individuals to make decisions by distributing authority along with information through the network
■ fostering and harnessing the power of shared awareness and thinking.

In our parlance, this means developing battle-wise decisionmaking; organizing command and control to make good use of more battle-wise decisionmakers; and building battle-wise teams. As figure 6–1 suggests, progress along any and all of these three axes can help a force meet the cognitive challenges of warfare.

Figure 6–1. Increasing Battle-Wisdom in Networked Warfare

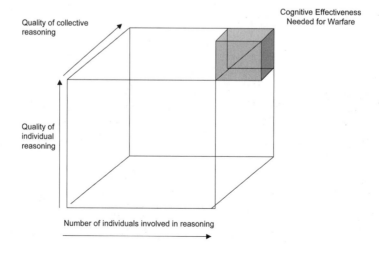

The Individual

Warfighters are expected to make sound and prompt decisions on matters of life and death, perhaps even war and peace, despite confusion, stress, and danger. The urgency and difficulty of making a decision may peak at inopportune times, such as under fire, when an enemy is in the target sights, or when well-laid plans go awry. Because of the wide variety of adversaries and contingencies associated with the new international security environment, warfighters have to make such decisions in unfamiliar and unpredictable circumstances. They must meld technical, operational, political, psychological, and moral factors. Moreover, because of networking, their decisionmaking must cope with more information and take into account interdependencies with other units and decisionmakers.

If one accepts Herbert Simon's observation about the difficulty of rational problem-solving, humans may not be up to such a complex challenge. However, our premise is that more, better, and shared information, because of networking, creates an unprecedented opportunity for warfighters to make better decisions. People can be better decisionmakers—more capable of anticipation, reaction speed, opportunism, and fast adaptation—provided that efforts are made to develop battle-wisdom in them and in how they make decisions.

If we are right that the ability to exploit networking will affect the course of future military competition and conflicts, the profile of the individual who can excel in wartime problem-solving, from seaman to admiral, is a matter of importance to national security. The mind of the networked individual must be good at receiving, assessing, and using information; analyzing data; deciding despite uncertainty; tolerating ambiguity; and sizing up unfamiliar situations. In short, the modern warfighter must be receptive, objective, and imaginative.

Such a description is not meant as an all-purpose personnel profile, applicable to all occupations with the military services. Functions such as supply, finance, training, and planning are as important as ever and can all benefit from IT, networking, and cognitive development. But these functions and the abilities needed to perform them are similar to those found in many knowledge-based enterprises. What is exceptional about the military realm is not the business side but the warfighting side—the requirement to reach urgent judgments in the midst of confusion and violence. We doubt that the need for battle-wise warfighters can be satisfied simply by recruiting, developing, and assigning well-rounded people of high but generic intelligence and ability. Rather, some fraction of military personnel should be optimized for battle-wise abilities, traits, and decisionmaking—much as the military selects certain types of individuals for special operations forces (SOF). In addition, because networking provides more and better information in flatter organizations, the performance of networked forces will depend on having as many battle-wise people as possible.

With growing volumes of networked information available to support complex, high-pressure problem-solving in many nonmilitary sectors and markets, competition will be fierce to find and retain people with mental abilities analogous to those encompassed by battle-wisdom. In the world of military networking, from superpower to second-tier powers to terrorist groups, those that value, find, develop, and effectively use such people will have an advantage over those that do not.

Networks help by ingesting, screening, sorting, posting, and feeding information. But it falls to humans to react to whatever information is furnished and make decisions only they can or should make. How can the individual, with the power of the network at his or her disposal, excel as decisionmaker to the point that the network creates operational and, in turn, strategic advantages? Simply stated, by *knowing and deciding.*

First, the individual should know what he or she is in the best position to decide. In principle, by having all significant and relevant information available on the network (assuming none is denied) *plus* tacit knowledge from local observation, the individual should understand his or her circumstances better than anyone else. Endowed with this information advantage, along with the requisite authority, he or she is in the best position to solve whatever problems surface. If the person is leader enough to take responsibility for the decision, and assuming a wrong decision will not endanger other units or their missions, little is to be gained and much potentially lost—notably, time—by referring the problem up the chain of command.

Second, once the network has done its part in preparing and presenting information, the individual must know how to use pulled information and blend it with tacit knowledge to solve complex problems with intuition and reasoning. Typically, this is a matter of identifying options and choosing among them—in the simplest case, choosing between doing nothing and doing whatever one can to avoid the penalty of doing nothing. Under intense time pressure, the tendency is to choose the path that intuition signals is good enough, or least unlikely to lead to disaster—not necessarily maximizing the chances of a favorable result, but at least reducing the chances of a regrettable one.

The cognitive abilities of particular value in networked warfare—anticipation, reaction speed, opportunism, and rapid adaptability—suggest a decision stream rather than a set of discrete choices. Anticipation is more likely to pay off if follow-up is quick. At the same time, the ability to decide and react quickly may give the warfighter more time to get and consider more information without losing the initiative. Opportunities can be created by anticipation and quick reaction, which place one in a position to see and seize unexpected chances for advantage. If the battle is not unfolding as hoped, the ability to adapt rapidly could be indispensable. Indeed, whatever perils are involved or mistakes made in anticipating, reacting quickly, and seizing opportunities can be remedied by rapid adaptability, thus reducing risk. This is the way the battle-wise warfighter thinks.

After reaching a decision, the individual must be willing and able to learn objectively from the consequences, good or bad, of choices and actions. The capacity to learn, thanks to both information from the network and the cognitive abilities of the battle-wise individual, is vital for the battle-wise decision method. A rapid learning-reasoning-acting-learning cycle can be hard to achieve under the best of circumstances. Under the dangerous and disorderly conditions of warfare, the cognitive challenge is immense. Adaptive decisionmaking in combat conditions, when time is scarce and bullets are flying, takes self-awareness, aptitude, training, and practice.

The ability of individuals to reason in networked warfare can be improved in three ways: raising the level of battle-wise abilities in the entire pool from which individuals are drawn; being more selective in finding and favoring individuals who have battle-wise abilities; and honing the abilities of those individuals.

Raising the level of battle-wisdom in the pool requires rethinking general military recruitment standards, screening, and priorities with an eye toward the mental abilities known to be important in solving complex problems in battle. Identifying people in the armed services with these qualities and getting them into warfighting positions could require changes or improvements in sorting, selection, promotion, assignment, and retention policies. Developing the problem-solving abilities of people in the force requires fine-tuning education, training, and career development programs to stress these abilities. All of this is explored in chapter eight.

Decentralizing to Involve More People in Decisionmaking

A network can mobilize individuals, each contributing to solving the problems and meeting the challenges facing a force during operations. The more people who are enabled and empowered to take advantage of network information to make informed local decisions, the more likely it is that the operation will succeed. In simple arithmetic terms, accompanying the distribution of information with the distribution of authority multiplies the number of effective problem-solvers. Depending on the average span of control, the combination of distributed information and authority can increase dramatically the amount of brainpower actively engaged. If that brainpower is battle-wise, and if clear but flexible rules govern who decides what, a force can gain marked operational advantages, especially in complex warfare.

In addition, the specific cognitive abilities that are crucial to operational success—anticipation, reaction speed, opportunism, and rapid adaptation—all strongly correlate with decentralization of authority. Precisely because combat success may depend on making time an ally—enhancing time-information—the case for decentralization is compelling. Yet distributing decision authority is counter to military tradition, in which authority drops off steeply the further down the pyramid of command (and rank) one goes. To gain strategic benefits from network technology and principles, military tradition will have to yield to progress.

Broadly speaking, the information and geopolitical revolutions are accelerating a trend toward decentralization and democratization of decisionmaking. From the time of ancient Greece, philosophers have debated whether to entrust authority to an enlightened *few*—perhaps to an omniscient *one*—or instead rely on the minds of the *many*. [1] The past half-century or so has tipped this debate decidedly in favor of the many. The failure and collapse of fascist and communist dictatorships in the 20th century and the corresponding supremacy and spread of liberal democracy have proven the fallacy of relying on the few to the exclusion of the many. Similarly in business, the advantage of empowering employees has gone from fad to fact of life. Under most conditions, common sense argues against counting on a chief executive officer or commander to be so brilliant that the reasoning of the well-informed many can be disregarded.

True, hierarchies still have certain indispensable functions: investing and allocating scarce resources; clarifying responsibilities; making strategic decisions; establishing standards, procedures, and policies; and providing insurance, checks, and balances against poor performance. When it comes to operations, however, creating options for decentralized decisionmaking is the key to agility, adaptability, and success in most enterprises.

Decentralization enables the best use of talent. In one sector after another, hierarchies have given way to horizontal trails of responsibility, decentralized decision processes, and democratic cultures. In most endeavors, it is more productive to allow individuals to use their talents to the fullest within broad guidance and flexible boundaries and encourage natural teams and networks to form than to rely on rigid job descriptions, a climate of strict dos and don'ts, and compartmentalized specialization. Especially in information-rich enterprises, the loss from constricting people usually exceeds any gain from preventing their mistakes. In any sphere, eschewing micromanagement in favor of distributed decisionmaking is a logical corollary of networking principles and a prerequisite of cognitive superiority.

The potential of networking in warfare cannot be realized unless and until command and control is reformed. There is burgeoning interest in how to fashion command and control systems for better results, given the demands imposed by the new security era and the opportunity presented by networking and the information that networks carry. Some have said that new command and control systems, as well as the forces they manage, should be resilient, flexible, responsive, innovative, and adaptive.[2] Amen. Decentralized command and control systems capitalize on cognition in information-rich networks and add flexibility in fluid situations.

For all the attention given lately to the advantages of flattening organizations, empowering subordinates, and jettisoning inflexible hierarchies, the U.S. military has some distance to go, whether compared to nonmilitary organizations or some non-American military organizations. The British, Canadians, and Australians, for example, have traditions of relying heavily on the judgment and initiative of junior officers, which they are now starting to draw on to exploit networking.[3] German military doctrine, even in the Nazi *Wehrmacht*, has followed the principle of issuing broad guidance and counting on officers in the field to make the right decisions.[4] To some extent, this is happening in the U.S. armed services—with seniors giving juniors the authority to make decisions, and juniors accepting and excelling with that authority.[5] Yet evidence abounds—many anecdotes coming from junior officers and cases witnessed by the authors—that senior U.S. combat commanders regard their newfound ability to observe the entire battlespace in detail as an irresistible opportunity to micro-manage.

At the same time, historical and current examples of decentralized authority exist and are well entrenched within the U.S. military. The success of U.S. forces in World War II has been attributed in part to the trust placed in subordinates.[6] The U.S. Navy, like all navies, could not function if ship captains were denied more or less full authority over what occurs onboard. As a result, delegation and individual accountability feature more prominently in naval culture than in army culture, where counting on buddies may be a higher value. Of course, the naval practice of delegating authority is motivated not by networking but by the autonomy of individual ships and the historical difficulty of communicating on the high seas. Still, with less reliance on hierarchy when operating, naval units and officers may find it easier to network—at least with other naval elements and with the Marines—than those of more centralized services. An even more striking example of autonomy and networking are SOF, who have been trained to collaborate at tactical levels with any unit of any service. Small SOF units led by junior officers or NCOs often are inserted or tasked to remote places. SOF are as battle-wise as any existing forces and accustomed to delegation of authority.

Compared to networked warfare, historical set-piece mechanized campaigns and battles required fewer and more discrete decisions; consequently, centralized decisionmaking was tolerable. Increasingly, fluid conditions and distributed enemies demand rapid and flexible operations and decisionmaking, which argue for locating authority "near the guns." Decentralization is especially important in contingencies where ambiguous and unfamiliar situations increase the value of tacit local knowledge.

At the same time, those at the top of a force must specialize and excel in what they are in the best position to ponder and decide, namely, the formulation and adaptation of strategy. Forces, networked or not, will perform poorly if, all else being equal, senior commanders fail to read overall patterns and express their intent accordingly. In addition to underutilizing the power of networking, commanders who succumb to the micromanagement temptation may give insufficient attention to strategic patterns, analyses, and choices. The aim is not so much to shift all problem-solving responsibility from seniors to juniors but to ensure that each tier is engaged in solving the sorts of problems it is best prepared and informed to solve.

The danger of micromanagement is aggravated by the fact that, in the U.S. military, *joint* command and control is shallow—it is mostly found at the joint task force command level and, to some extent, with land, sea, and air component commanders. Technically, networking enables operational integration as deep as communications permit, and integration can provide enormous leverage. In practice, however, the absence of deeper joint command and control will mean that force and component commanders will be overly involved in small operations. This will deny the full benefits of networking and retard progress toward integration. Pushing joint command and control down deeper into the force—the details of which are immensely complex and beyond our scope—is an important task in the strategy to build and field battle-wise forces.[7]

Engaging the battle-wise decisionmaking talent of more individuals in a networked force is not just a matter of devising command and control architectures to shift authority and initiative downward and outward. As noted earlier, networking not only informs individuals and units but also makes them interdependent. New command and control arrangements should facilitate peer-to-peer collaboration so that units throughout the network can support and be supported by one another regardless of geography, service identity, and formal operational command boundaries. Decentralization to involve many in local problem-solving may weaken vertical control but strengthen horizontal and diagonal links, which are far harder to fit with rank and rigid structure. Even if top commanders share much of their authority, they have a role in facilitating such links and mediating interdependencies—all the more reason to eschew micrmanagement.

Figure 6–2. Warrior-Centric Network

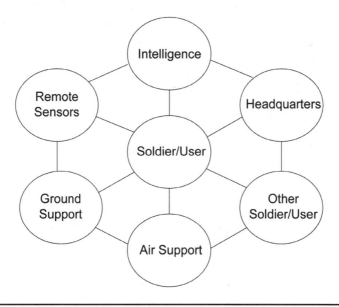

The power of distributing information and authority argues for a conceptual shift from network-centric warfare to warrior-centric network, as depicted in figure 6–2. This, after all, is the way the Internet and other networks work, and it is the way warfighters and the Armed Forces should think of their operating networks. In the words of the U.S. Chief of Naval Operations, "The sailor is in the center of the combat system and not an appendage." It is not hyperbole to say that all warfare becomes local when a relevant operating picture can be constituted at any location. In the mind's eye, the junior decisionmaker, having migrated from the bottom to the edge of the force, must now move from the edge to the center of his or her network.

Figure 6–2 underscores the principle that the truest measure of any network's worth is the level of satisfaction of the needs of its user. It also implies that flexible collaboration among warriors is an option of growing importance, not available when communications could only conform to and reinforce vertical command and control. Figure 6–2 makes explicit that the highest power of information is the enablement of the person and, thus, illustrates why the cognitive abilities of every user are crucial. It reminds us that data should move by the force of smart pull.

One of the chief reasons that the cognitive abilities of small-unit leaders are becoming so important is that networking permits forces to be distributed in small but connected units, which is advantageous for maneuverability, survivability, and effectiveness. This means that more junior officers and NCOs must be wise enough to make decisions that people of their rank were rarely called on to make in the past. It also means that the intuitive component of problem-solving may be less reliable, since people in the rank-and-file are less experienced than those at the top. More individuals will require better cognitive abilities to solve more and different types of problems, with less ability to draw on past experience because military operations are diverse and changing.

Collective Intelligence

In the military and other fields, decisions are seldom crisp choices made by "lonely" individuals at the top. Indeed, by mobilizing the many, networking works against single-point decisionmaking. In wired organizations, decisions can and often do reflect shared analysis and judgment. Think of corporations whose success can be attributed to their collective wisdom. Google has become legendary for pooling the intelligence of its people. The resurgence of General Electric during the information age can be attributed as much to the learning-sharing-thinking power of the entire culture as to its incandescent ex-chairman, "Neutron Jack" Welch.[8] The genius of Wal-Mart, a favorite of business school case studies, lies in how its logistics processes are managed and improved by smart teams, not by individual wizards.

In *The Wisdom of Crowds*, James Surowiecki explains how and why groups consistently produce decisions and solutions superior to those produced by all but a few of the individuals in the group. The most consistently accurate way to guess the number of jelly beans in a jar is to ask a large number of people their estimates and then take the average. The best way to set odds on a football game is to let a large population of bettors do it. The fairest way to decide the fate of an accused is by jury.

The reason for this, simply put, is that the errors of individuals tend to cancel out one another as numbers increase, leaving the average to be that much better. But such collective wisdom only works if there is ample diversity and independence of views among the participants. That way, the full range of experiences, perspectives, and information of the many are in play, resulting in a better answer than if the solution is based on the experience, perspectives, and information of only a few, even if they are of above average intelligence. Absent diversity and independence, "groupthink"—the foe of reason—may emerge.

Enhancing collective intelligence in the military is bound to be more difficult and perilous than in less disciplined organizations and less dangerous domains. Translating Surowiecki's requirement for diversity and independence into military forces and operations would, to say the least, require a great deal of consideration and care.[9] Urgency and violence tend to lead to single-point (though not necessarily centralized) decisionmaking, and rightly so. Moreover, military units are highly structured; decisionmaking tends to conform to structure; and structure impedes collective thinking. Military organizations also are no longer fixed once operations begin; instead, they undergo continuous change in response to conditions, especially in joint operations. A frigate may begin as part of a carrier battlegroup, move to providing firepower to an amphibious strike force, go on to link up with SOF searching for an enemy command center, and later join other frigates to patrol a coast. This obviously complicates the applicability and realization of collective battle-wisdom.

Yet because networking opens horizontal paths for collaboration, individuals must be able to reason together, from pairs to small teams to entire forces. Shared awareness allows for the possibility of shared problem-solving. We all know that solving problems by committee can be frustrating, time-consuming, and fruitless. Moreover, groupthink—the opposite of diversity and independence—is especially strong in military cultures and can hurt instead of help performance.

As networking facilitates integrated operations, large traditional military structures—corps and divisions, air wings, carrier battlegroups—begin to lose their utility and integrity. In time, networks themselves could become operating structures—though ever-changing ones. According to networking theory, more or less confirmed by practice, ad hoc teams will self-organize to deal with common problems, enabling a force as a whole continuously to optimize its resources despite uncertainty and change. Such fluid organizations will work only if awareness, common intent, and opportunity for collaboration are shared, which networking can provide.

A commander soliciting the independent and diverse opinions of staff or subordinate commanders before making a decision is not a novel concept. Yet the commander will determine what weight to place on the views of others as well as his or her own—and may be wrong. An interesting twist, being tried in some corporations, is to go with the consensus whether the commander sees it the same way or not. Apart from the obvious time-consuming drawback of this method, commanders are taught, for good reason, that they must take ultimate responsibility for failure and cannot do so if the decision is collective. Of course, the commander with confidence in the wisdom of subordinates may elect to take responsibility for the method and the consequences of team decisionmaking.

While collaboration is better than solo problem-solving in theory, it may not be that simple in practice. A balance must be found between the speed and agility provided by single-point decisionmaking and the quality of decisions based on the independent and diverse views of many. When considering what kind of problem-solving is best to maximize battle-wise performance—anticipation, reaction speed, opportunism, and fast adaptation—the answer must be: it depends.

Notes

[1] A treatment of this debate in Athens is found in I.F. Stone, *The Trial of Socrates* (New York: Random House, 1989).

[2] David S. Alberts and Richard E. Hayes, *Power to the Edge: Command and Control in the Information Age* (Vienna, VA: CCRP, 2004).

[3] The British military prefers to speak of "network-enhanced capabilities" instead of "network-centric warfare" because the latter connotes the centrality of the network instead of its users.

[4] Alberts and Hayes.

[5] Dan Baum, "Battle Lessons," *The New Yorker*, January 17, 2005, 42–48.

[6] Stephen E. Ambrose argues in *Citizen Soldiers: The U.S. Army from the Normandy Beaches to the Bulge to the Surrender of Germany, June 7, 1944–May 7, 1945* (New York: Simon and Shuster, 1997) that one of the reasons that U.S. Soldiers performed so well in World War II is that subordinates easily

took the place of leaders who were hurt or killed in combat. Thus, there was little loss in leadership, continuity, or effectiveness when those in charge were knocked out of action. In contrast, German soldiers had a much harder time fighting effectively when their leaders were lost.

[7] For more details, see Alberts and Hayes, *Power to the Edge*, and David S. Alberts, *Understanding Information Age Warfare* (Washington, DC: CCRP, 2001).

[8] The clearest accounts of this metamorphosis of General Electric come from Welch himself, who explains how his role changed from commander to enabler as the company became a more networked, learning organization. John A. Byrne, "How Jack Welch Runs GE: A Close-Up Look at How America's #1 Manager Runs GE," *Business Week Online*, June 8, 1998, available at <www.businessweek.com/1998/23/b3581001.htm>.

[9] James Surowiecki does not call for the application of collective decisionmaking to the military field or even address the matter in *The Wisdom of Crowds: Why the Many are Smarter than the Few and How Collective Wisdom Shapes Business, Economies, Societies, and Nations* (New York: Random House, 2004).

Chapter Seven

An Excursion beyond the Military

Why Observe Broader Research and Experience?

Although U.S. Armed Forces are already made up of intelligent, dedicated, and skilled men and women, imagine a force of warfighters finely tuned specifically to excel in networked warfare—conditioned to think and decide using a blend of reliable intuition and quick reasoning and unmatched in their ability to anticipate, decide promptly, capitalize on opportunities, and learn in action. In addition, the soldiers of such a battle-wise force would know when and how to leverage the collective intelligence of operating teams and how to pull information from their networks to enhance decisionmaking when their own mental models and experience fall short.

Achieving this vision will take a purposeful strategy. To treat the pursuit of advantage on the cognitive plane as just another facet of developing network-centric capabilities—as the U.S. defense establishment is doing—is to underestimate the special nature of the challenge. Although the military has unique mission structures and needs, it should, as part of designing such a strategy, explore a wide range of research, including that which is not aimed at the Armed Forces and warfare. It also should study how other sectors and institutions are trying to develop, organize, and use individuals and teams with comparable abilities and traits. As with other facets of exploiting and adjusting to IT and networking, other organizations have been struggling with similar cognitive and personnel challenges longer than the military, forced to do so by their recognition that how and how well people think affect the bottom line in the information age more than ever.

Of particular interest is research and experience outside the military sector in relation to the use of key cognitive abilities for operating in urgent, high-stakes, complex, fluid, and information-rich situations. While keeping the unique aspects of military duty in mind, much can be learned from companies that have successfully implemented such practices as rigorous hiring of top talents, self-directed learning systems, leveraging of collective intelligence through encouraging decentralized decisionmaking, using cross-boundary teams, cross-training, inducing artificial disruption in a system's operations, and fostering collaborative decisionmaking.

The Individual and Intuition

What is known about intuition comes from observations of decisionmaking in a wide variety of settings. Modern research, of which the work of Gary Klein stands out for its clarity and practicality, removes some of the mystery about intuition.[1] Klein's work is based on a person's experiences and subsequent understanding of complex situations drawn from repeatedly seeing patterns, making decisions, and receiving timely feedback from those decisions. Intuition is essentially educated anticipation of how a situation will unfold. Intuitive decisionmaking comes naturally to the experienced warrior, businessperson, athlete, or first responder. Rarely do these decisionmakers attribute their decisions in urgent situations to the weighing of options or to rational deduction involving the assessment of a problem using a deliberate process or scientific method. About 90 percent of the time, in fact, people rely on an intuition-based process to make urgent, important decisions in fields as diverse as nursing, athletics, firefighting, weather forecasting, and business.[2] The standard answer to urgency is to size up the situation, set a direction, and act.[3]

Intuition enables such a quick cognitive response based on a combination of immediate, largely tacit knowledge and preexisting mental models. We hold beliefs about how certain processes and phenomena work, and we form models based on those beliefs. Expert firefighters know how fires spread, the flammability and reaction of different materials to heat and flame, and the behavior of different types of fires in different conditions. Nurses have an understanding of how infections start and spread, what a healthy person looks like, and what symptoms imply what diseases. Coaches comprehend the nuances of the sport, communication techniques for motivating players, and the skills needed for different positions. Such understanding of the critical elements in a field becomes a filter through which to interpret new information, unconsciously predict the outcome of a course of action, and decide.

Experienced firefighters or emergency rescue workers, for example, have well-developed mental models after seeing a plethora of scenarios played out in the course of their work. Rarely do they analyze and compare multiple options when under time pressure. Rather, they project mental simulations of what might happen with a particular course of action because they know how similar situations have unfolded in the past. If their initial response to a stimulus is incorrect, they often have the objectivity, flexibility, and experience to correct their course accordingly in a rapid and iterative manner.

Firefighting is one field where intuitive decisions are regularly made under the extreme time pressure of putting out a dynamic and dangerous fire. Veteran firefighters can recognize situational patterns rapidly and usually develop a plan of attack within moments. In one instance, a fire chief fighting a house fire made a quick decision to leave a house after several failed attempts to fight flare-ups in the first floor kitchen. The chief attributed his decision to extrasensory perception, but in-depth interviewing later revealed that he perceived a subtle difference in the sound and heat intensity of the fire. He had seen, heard, and felt so many fires in his life that he was able to recognize immediately that something was missing from the usual patterns—even though he did not consciously understand exactly what was missing at the time. His intuition told him something was different, and he made the call to exit the house moments before the floor collapsed and revealed a raging fire in the basement below the living room where, moments before, his crew had stood. The chief had enough understanding of fire behavior to recognize an anomaly, and he correctly assessed that the situation was more dangerous than normal.[4]

Because intuition is not an analytical function, we may not need complete information, logic, or reasoning to recognize patterns and make good enough decisions in a complex situation—assuming we have seen a similar situation before and our mental models enable recognition of at least some of the situational patterns. As Malcolm Gladwell suggests in *Blink,* seasoned people in many occupations are able to "thin-slice" information—scanning, consciously or unconsciously, for a few key indicators or predictors instead of trying to absorb myriad information, as less practiced people must do before reaching a judgment. To those who know what nuggets to look for, the time required to survey a "thick slice"of information is not worth it. Conversely, when time is not available, the ability to thin-slice may be invaluable.

Expert rock climbers do not necessarily know how to negotiate a section of rock face until they are actually within a few feet of it and can see details, such as pitch, handholds, cracks, and anchoring points. With a fairly quick look from the ground, however, they can judge their ability to climb the entire route, and decide on a general direction to go. They then project mental simulations of each move as they climb upward, with each step revealing the next. They respond to problems as they encounter them, one move at a time, relying on their intuition without having to reason. Because they have a memory full of experience, they recognize patterns in rock faces that they have not climbed before and they can respond intuitively with appropriate moves as they go.

An expert gymnast, despite sharing some physical and mental traits with a rock climber—flexibility, strength, balance, muscle control, focus, and determination—may not have the mental models to climb a challenging rock face without training in the disciplines of rock-climbing. Intuitive decisionmaking is different from random or lucky guessing because the decisionmaker has a higher probability of making a correct choice than one who may have never faced a certain type of situation. A life spent training for soccer does not prepare an athlete to excel at American football without practicing the techniques and patterns specific to the game. Although an athlete from one sport might master another more quickly than a non-athlete, immediate cognition is a skill that goes beyond guessing. An individual must learn the types of patterns in a particular field of operations to sustain successful results.

Reliance on intuition does not mean acting impulsively without regard for fresh information. Indeed, one must be open and sensitive to information that quickly reveals patterns in a dynamic environment. At the same time, applying immediate knowledge without a relevant mental model might be unreliable and even dangerous in situations with high stakes, urgency, and complexity. It also helps to know the limits of one's experience, as well as when to seek additional information or advice when mental models are irrelevant to the problem at hand.

Relying exclusively on intuitive decisionmaking might be adequate when the stakes are not so high, the problems are repetitive, or a lack of time leaves no alternative. Whether in military or nonmilitary activities, however, intuition alone will not suffice when circumstances and problems are complex, fluid, and unfamiliar. Moreover, networking technologies, structures, and tools provide an unprecedented opportunity to strengthen the reasoning component of decisionmaking, even when time is scarce. Individuals and organizations now can fill in the gaps where intuition falls short, and replace, verify, or change intuitive leanings.

Combining Intuition with Reasoning in Practice

Again, decisionmakers must understand the bounds of their mental models if they are to utilize intuition and reasoning synergistically. In addition, they must have a high degree of objectivity in assessing a situation to determine if their mental models are applicable. Situational complexity is a function of the number of parameters involved in the scenario at hand.[5] The greater the number of parameters in a scenario, the greater the number of outcomes possible, and the less intuition applies.

Albert Einstein once said, "Our theories determine what we measure." Likewise, our mental models—cognitive constructs that translate into beliefs, assumptions, rules of thumb, biases, and theories as to how the world operates—determine the way in which we see or diagnose a problem. Two individuals may look at the same reality but see two different combinations of issues because of differing mental models and experiences. A perfect appreciation of one's mental models would require greater objectivity and detachment than most of us have. However, individuals with a higher degree of self-awareness and facility with reasoning under pressure will have an advantage in making intelligent decisions when it counts.

Neonatal intensive care and the forecasting of violent weather stand out as two fields in which decisionmakers have been able to combine their intuitive abilities with analysis, tools, and reasoning to enhance their decisions.

In intensive-care wards for premature babies, experienced nurses know how to detect the presence of an infection called sepsis that premature babies can develop. If not detected immediately, sepsis can lead to a rapid death. Diagnosing sepsis involves a combination of factors, including visual cues that instrumentation or nurses who have never witnessed the symptoms cannot easily detect without understanding the whole picture. Any one of the indicators, such as change in skin hue, elevated temperature, lethargy, and swelling might appear on its own without much consequence; these symptoms occur frequently in both healthy and premature newborns. However, the combination of all the symptoms at the same time is what one must recognize through developing appropriate mental models.[6]

In one instance, a newly hired nurse on duty failed to notice the combination of cues indicating sepsis in one of the babies in her ward. Fortunately, her supervisor happened to walk by and noticed that something "didn't look good" about the baby— without ever before having seen that particular baby. The supervisor suspected sepsis. She quickly used available information from various instruments to check her intuition. The baby was treated with antibiotics and then given a blood-culture test to confirm the visual diagnosis. The blood-culture test proved the nurse's intuition correct 24 hours later.

The veteran nurse knew she had to act promptly and on the basis of incomplete information. She also knew that if her intuition and initial action proved wrong, the alternative causes of the symptoms would not have led to rapid death. In contrast, the new nurse was relying on instrument information alone.[7] Self-aware decisionmakers in urgent situations often rely on their intuition first, and then confirm or correct their intuitive response with available data as time permits. The less familiar they are with the problem they face, the more they must rely on analysis of all relevant information—pulled from information systems and networks—before acting.

Intuition and analysis also combine in the forecasting of violent, sudden weather, another field where problems can be urgent, complex, and destructive. The best weather forecasters tend to rely on their intuition first before using instruments or turning to the analysis of others. They check the dew on the handrails as they walk out the door in the morning, notice the impact of their footprints in the grass, feel the temperature of the air, and, of course, look at the sky. They assess the whole picture and then build mental models from their accumulated understanding of patterns and correlations of different factors. When they arrive at work, they then check the most recent raw data themselves rather than relying on the interpretation of others. This enables them to see trends, patterns, key indicators, and anomalies in the data—somewhat like the way the firefighter noticed differences in sound and heat that cued him to leave the burning building. Weather forecasters know how to look for more data where needed, build mental simulations of what might be unfolding in the weather, develop a story with the patterns they are seeing, compare their intuitive findings with the data, adapt as necessary, and make a judgment when they feel they understand current reality.[8]

With the advent of networking, human beings have an opportunity to leverage IT for enhancing and verifying intuition, as well as supplementing intuition when additional information and reasoning are needed to make a good decision. If the firefighter mentioned above had waited for analysis to confirm his intuition, he may not have gotten his crew out before the floor collapsed. On the other hand, if the firefighter had had access to a networked sensor or camera in the basement of the burning building, he might have been able to make a better decision on where to attack the fire and have avoided the risk altogether. If the neonatal intensive care nurse had waited 24 hours for test results to confirm her hunch about sepsis, the baby's infection might have been fatal. However, if the monitoring equipment had not confirmed the independent symptoms of sepsis, the nurse might not have been as confident in her immediate response.

It seems that decisionmaking in high-stakes, urgent, complex situations where sensors and networks of information are available will require a blend of well-developed and relevant intuitive abilities, awareness of the boundaries of mental models, and the ability to analyze data and reason quickly.

Demands of the marketplace also are becoming more urgent, fluid, and unfamiliar as information technologies change the dynamics and pace of reality. Those same technologies, especially when networked, can provide information that can check and buttress intuition, strengthen reasoning, and combine intuition and reasoning in improved decisionmaking. Finding and developing individuals who can excel at the intersection of abundant data and turmoil—akin to the information and geostrategic revolutions described earlier—will become increasingly important for an organization's advantage. Where and how are such people found?

Recruiting Intelligent Decisionmakers

A study of 11 companies that have consistently outperformed the market for 15 years found that their managers tend to emphasize hiring the right people, even before corporate strategy, vision, or technology. By hiring focused, intelligent, versatile people, they create a culture of discipline, learning, flexibility, and adaptability—all of which are necessary to thrive in a complex and dynamic environment.[9]

Google aims to hire the top software people in the world. The company operates in a highly competitive and dynamic environment where technology changes rapidly and competitors are fighting for a foothold in a saturated market. Google selects people who not only are superlative software engineers but also have the entrepreneurial spirit needed to take risks, to "fail early" before a decision goes too far down the wrong path, and to troubleshoot on the fly. Despite its size, Google is extremely selective: it hires roughly 1 out of 1,000 applicants.[10] Lately, it has managed to hire a majority of the best available search-engine people in the world. Such high standards and rigorous discipline in hiring give the company an enormous amount of trust in its employees, enough for them to authorize and encourage distributed problem-solving without approval from on high. Consequently, decisions are made more quickly when responding to problems.[11] As testament to its approach, Google recently went public and raised $1.7 billion—one of the largest technology initial public offerings ever.[12]

Wells Fargo is another company that emphasizes people above all else. With a hunch that big changes were coming to the banking industry in the late 1970s, the chief executive officer of Wells Fargo hired some of the most talented management teams in the industry. Rather than trying to predict what the changes would be and gamble on a strategy, he focused on building a team of the best minds he could find. As a result, Wells Fargo survived the banking deregulation and shakeout of the 1980s. It outperformed the stock market by a factor of three, while the banking sector as a whole fell way behind.[13]

Other evidence points to the fact that hiring people with appropriate qualities for a job is of utmost importance. From a poll of about 80,000 managers from 400 companies, Gallup concluded that each human's nature and talents are unique and that people do not really change their behavior that much.[14] For the most part, corporate education and training or a sweeping cultural change are unlikely to change or set people's attitudes, talents, or motivation levels. Like a good National Football League football coach choosing new players, the best managers draft for talent and attitude, and then assign people responsibilities where they can become more and more of who they already are.[15]

A good example of this can be seen at Southwest Airlines, the most profitable airline in the last 20 years. The Southwest philosophy is to hire people who are innovative, self-confident, and fearless about finding better ways to solve problems. Southwest hires people who have an attitude that will fit its culture and then trains them to develop the skills for particular jobs.[16] Because Southwest is clear about its values and purpose, it does not waste time trying to alter attitudes that do not fit. Abbot, Circuit City, Fannie Mae, Gillette, Kimberly-Clark, Kroger, Nucor, Philip Morris, Pitney Bowes, Walgreens, and Wells Fargo—all of which have outdone their respective markets for 15 consecutive years—practice a similar philosophy and make hiring people that share their values and purpose a priority.[17]

Although the exact qualities for which such people-first companies hire may differ, some common traits are necessary for operating in an increasingly complex and dynamic marketplace. The list includes creativity or unorthodox thinking; the ability to thrive in ambiguity, complexity, and pressure; willingness to learn and change in action; advanced reasoning capabilities, both intuitive and analytical development; self-awareness and objectivity; measured confidence; and a willingness to take responsibility and be accountable for the consequences of decisions.[18] In short, organizations want people who are willing and able to make intelligent decisions with incomplete information and under pressure.

Despite the glitz of IT, and because of the ubiquity of information, differentiation among organizations will come down to the people who know how to exploit information. High-tech companies like Google do not want good engineers, but rather great engineers who also can innovate and initiate. As the complexity and speed of markets increase, the companies that win in their markets and in stock markets will be those full of people who not only know how to use the tools of their trade but also are capable of knowing how to blend reasoning and intuition when making decisions. Companies with a proven record will find it is easier to compete for such people, with both compensation and reputation. However, attracting and hiring talented people requires that the highest priority be placed on finding and keeping such people.

In sum, a sampling of strong companies reveals that they all believe that people are of unsurpassed importance; that cognitive abilities and the inclination to innovate, initiate, and take responsibility in fluid markets, utilizing the power of information, are essential attributes; and that the key to having such people is to find and hire them from the start. We will consider later whether this formula is right for the Armed Forces.

Improving Decisionmaking

In both civilian and military domains, we have observed that because intuition springs from experience it can be less reliable as rapid change occurs and problems become less repetitive and familiar. Mental models are less likely to aid in comprehension if reality is fluid, messy, and unpredictable. Yet developing intuition is possible. Moreover, because it is normally the initial response in urgent situations, it should get attention in improving decisionmaking.

Years of seasoning and repetition traditionally have been essential in building reliable intuition in veteran firefighters, nurses, weather forecasters, and the like. But understanding how intuition works and consciously practicing decisionmaking can offset inexperience to some degree. Understanding and practice can have two positive effects: shortening the time it takes to develop good intuitive tools and bolstering the reliability of intuition in unfamiliar circumstances through developing more attuned self-awareness.

Intuition training programs already have been adopted by fire departments, the National Fire Academy, business executive training programs, and parts of the Armed Forces. The method is straightforward: isolate the *sorts of* decisions (as distinct from specific decisions) that one could come across in certain *types of* situations (as distinct from specific situations) given the job and the environment; practice those decisions repeatedly through real-life situations or simulations; and review the results of the decisions to learn and adjust. This is a much more active approach than the occasional training seminars or certification testing that many companies already do. It treats a job as a discipline and is similar to the approach an athlete or musician might take in training.[19]

A corporate executive has to make many critical decisions, such as setting budgets, selecting contractors, identifying opportunities for investment, hiring and promotion, and assessing the viability of a project.[20] A soccer player might identify and practice such decisions as passing or dribbling, shooting the ball or passing when in front of the goal, playing zone or man-to-man defense, positioning on the field in relation to the ball, or even how hard to kick the ball based on field conditions. A commercial airline pilot might actively practice decisions, such as when to climb out of turbulence, or even how to recognize extremely dangerous weather scenarios, and when to change course. A police officer might practice decisions such as how to recognize when the use of force is justified and when it might be necessary to call for backup.

Analyzing the types of decisions an individual makes in a job can reveal a better awareness of mental models. Such analysis can indicate what makes decisions difficult, cases where mental models do not apply, and potential pitfalls or habits that might lead to failure when it counts. As a novice, some decisions—as in the case of a junior business executive or pilot—might require more analysis of data. By repeatedly exercising representative decisions and assessing their results, intuition should become both deeper and more reliable, and the requirement for time-consuming deliberation and analysis can be reduced.

Accelerating the rate at which experience and sound mental models are developed should improve the intuitive component of decisionmaking. However, training cannot and should not try to anticipate *in detail* the variety and ambiguities of dynamic markets. Therefore, development of reasoning skills is critically important for making good decisions. Methods of improving reasoning skills are not that different from the intuition-building processes just described. The major difference is that rational decisionmaking is more structured—that is to say, one should follow more or less standard conscious steps, such as defining the problem, gathering and screening information, clarifying objectives, disaggregating the problem for analysis, and identifying the best options.[21]

Once a problem is defined, three modes of reasoning are useful for decisionmakers: inductive, deductive, and abductive. Inductive reasoning involves the inference of a hypotheses based on the gathering of evidence; it moves from a level of specificity to a more general conclusion. Deductive reasoning seeks to identify data that confirms the truth of a hypothesis, moving from a general assumption to particular evidence. Abductive reasoning involves the use of analogy, whereby alternative or creative hypotheses are applied where scant evidence may exist.[22] All three types of reasoning should be understood and developed for solving problems. The crucial first step in any reasoning process—problem definition—will, in large part, determine what type of reasoning is used. Efforts to sharpen rational decisionmaking follow more or less the same pattern as intuition-development: identify what types of decisions one is likely to face, practice making them, receive feedback, learn, and repeat.

Whether trying to improve intuitive or rational decisionmaking, choosing what types of decisions to practice is vitally important. For many professions, creating a potent mixture of urgency, stakes, complexity, and flux is more important than guessing the specifics of real situations that may be encountered. In such circumstances, training should tax and build intuition, reasoning, and the self-awareness to blend the two. Again, this is the approach demanded by the combination of unfamiliar problems, abundant information, lack of time, and severe penalties for being either wrong or too slow.

Imagine a young wilderness fire commander—not dissimilar to the major of the ambushed peacekeeping unit discussed in chapter five—facing a situation out of his realm of experience. A powerful windstorm is moving in and will blow a wildfire dangerously close to some homes. The crew is on the opposite side of the fire from the endangered homes, and it is impossible for the entire crew to mobilize in time to protect them. In desperation and for lack of experience, the commander begins calling around to more experienced commanders to ask for advice. He has an assistant quickly assess the likelihood of mobilizing to arrive at the houses in time given the speed of the approaching storm. He has another assistant monitor the weather. Another assistant is making calls to headquarters to coordinate an evacuation using local volunteer and media networks.

Several recommendations are made by other fire commanders, and the local commander runs them by his assistants. One suggestion seems particularly interesting—to light another fire on the side where the crew is and use the power of the wind to create a backdraft that will suck the oxygen from the current blaze. The commander quickly does a network search and finds several articles written about backdraft attempts. He does a text-search for key words, such as *success, probability,* and *conditions* and is able to see that the conditions seem favorable for this technique and confirm the advice given by the other commander. He calls a few of the other commanders to ask their advice on this line of reasoning and although one disagrees, the other three think it is the best option and offer insight on problems that might arise.

Meanwhile, calculations are done to assess the probability of the backdraft technique, using velocity of windspeed and other factors, such as the chance that the wind might change direction and turn on the crew or spread to a group of homes to the south instead. Because the approaching storm is intense, the decision is made to dispatch half of the crew to mobilize volunteer firemen and homeowners to dig firelines and set up defenses for the homes. Finally, the commander takes a step back and considers the cost-benefit of saving the homes versus the risk to the lives of his crew and others in the direct line of the quickly spreading fire. Assuming the commander has an awareness of his mental models, is adept at defining problems, has the ability to communicate with others and seek diverse opinions, and has access to a network of pertinent information and the ability to run relevant analysis, this entire process would probably take less time and perhaps be of less consequence than choosing a faulty path based on a hasty decision outside the realm of his expertise.

The role of time and information and the importance of time-information in this hypothetical case are worth noting. The fire commander tries not merely to optimize the tradeoff between time and information but to expand time-information—using time to gain critical information that in turn permits a decision that is both timely and sound. Lacking experience, and therefore appropriately doubtful of his intuition, the commander borrows intuition from more experienced colleagues on the network and combines this with time-efficient analysis. What the young commander lacks in background he makes up for in self-awareness—as well as a good mix of humility and confidence.

The development of measured confidence to reason under pressure is important for decisionmakers in complex situations. This requires a degree of self-reliance, which can be developed through, among other practices, self-directed learning.

Self-Directed Learning

In addition to selecting the right people and rigorously training intuition and reasoning, organizations can provide the climate and tools to promote individual learning to improve the use of information and decisionmaking. Self-directed learning encourages and equips individuals who are able and willing to take initiative, adapt continuously, and make decisions without close supervision. Some of the most successful organizations in today's increasingly complex and dynamic world and markets are those whose people learn on their own, without asking permission or being directed to do so.[23] In effect, people smart-pull the training they need as they need it from distributed learning networks.

Southwest Airlines regards its support of employee learning to be an essential component of its competitive advantage. "Employees who embrace learning as a life-long pursuit are more alert, better informed, and more creative. This translates into new ways to simplify operations and cut costs, and new ways to better serve customers."[24] Research suggests that a distributed and self-managed learning model is more effective than traditional learning models for an organization facing an unstable environment. By nature and work, self-directed learners take prudent risks, are confident yet humble, are self-reflective yet careful listeners, have voracious appetites for information and ideas, and are open to criticism and change. The more unstable the environment, the more important it is for such attributes. Empirically, self-directed learning and high performance in jobs that involve a lot of change show a correlation.[25]

The self-directed, or smart-pull, learning model is more effective than an others-directed model because it enables greater relevance to the individual's needs, greater flexibility in the learning process and tempo, immediate and long-range payoffs for developing problem-solving skills, highly focused learning, and lower costs for an organization.[26] Of course, effective self-directed learning requires more than permissive corporate policy. The responsibility of the organization is to supply the tools, time, and incentives for an individual to pull the information, participate in decisions, and receive timely feedback from decisions.[27] Companies that promote self-directed learning, including Motorola, Honda, and General Electric, have reduced by up to 50 percent the cycle time for new product introduction and increased their market shares.[28]

By combining self-directed learning with training of intuition and reasoning, an organization can enhance the ability of its personnel to make good decisions. Obviously, self-directed learning fits well with networking, and it actually can help in preparing for real-life conditions in which networked users will be responsible for initiating smart pull and prioritizing information from the network.

Decentralized Decisionmaking

The value of self-directed learning can be undermined if individuals lack the trust and confidence of their superiors and are not granted the authority to make decisions. Many strong businesses are distributing decisionmaking authority to those on the front lines, a practice that not only enables an organization to act with greater agility and speed but also imparts confidence to those who make the decisions.

Businesses in the 1980s and 1990s were swamped with new management theories—to name a few: total quality management, continuous improvement, right-sizing, core competence, process engineering, strategic alliances, competitive strategies, learning organizations, empowerment, flattening of hierarchies, cross-boundary teaming, and even destroy-your-business (so you can build it anew). To our knowledge, none of these theories alone induced sustainable organizational change without the mutual commitment of leadership and rank-and-file employees.

For reform to be sustainable, an organization must put into practice certain values and principles concerning information, people, and trust: transparency; open information-sharing; cross-boundary communication and collaboration; an understood mission and values; a culture that rewards taking responsibility; a commitment to learning; and a willingness to give talent room and to give people the confidence and authority to make decisions.[29] Many companies, large and small, have achieved success by applying these principles and practices. Although we focus on decentralized decisionmaking, other ways of enhancing and harnessing information, people, and trust also can contribute to the overall success of an organization.

One model for successful decentralized decisionmaking in an organization is Semco, a manufacturing conglomerate in Brazil. Semco is a rather democratic firm that relies heavily on individuals at all levels across the company to make important decisions, thus strengthening adaptability in the face of change. For this type of organization to succeed, flow of information is key. For information flow to be open and transparent, trust must exist. Trust develops through peer-to-peer relationships. Trust has a higher potential of developing in an open and democratic corporate environment, where transparency and sharing information are the lifeblood of a culture committed to adaptability. Such an organization becomes more fit for responding to opportunities quickly with a nimble workforce unencumbered by bureaucratic rules and centralized decisionmaking.[30] Since adopting these practices in the late 1980s, Semco revenues have quadrupled, and the staff has grown from 450 to 3,000. The firm runs eight businesses, having expanded into outsourcing management (for four of Brazil's five biggest banks), environmental-site remediation, and engineering-risk management. Employee turnover rate is an astonishingly low 1 percent.[31]

Google, once again, stands as an example of the robustness, efficiency, and market success that distributed decisionmaking can foster. The company culled out managers in the engineering departments and instead has independent teams of three engineers who operate autonomously and fix whatever problems they see with no need for permission from above. Strict hiring practices save unnecessary supervision because employees can be trusted to do their jobs and make intelligent decisions. This approach also encourages risk-taking and creativity, both of which are important for a company to continue growing in increasingly competitive markets.

As predicted by theories of complex adaptive systems, corporate decentralization seems to be a rewarding way to function in a dynamic marketplace. The mythical superhero leader at the center with all the answers no longer exists; problems are too complex and markets too urgent for the one or the few to understand, decide, and act. Organizations that need to wait for bureaucratic procedures, chain-of-command review, or decisions from on high before acting on an opportunity may not be able to survive in fluid and unfamiliar situations.

Tapping Collective Intelligence

Even as organizations decentralize decisionmaking, enabling many individual decisionmakers to address many different problems, they also are pursuing the idea of collective knowledge and collaborative decisionmaking. The use of horizontal communication techniques, cross-training, and cross-functional teams can improve the quality of decisions and overall adaptability. Dialogue and openness, versus closed environments that breed stuffiness and defensiveness, enable a diversity of viewpoints to be voiced when addressing a problem. By letting each individual know how others in the company are employed and what they know, cross-training encourages empathy, builds trust, and lowers defensiveness—all crucial for collective thinking. It also enables a broader perspective on how all the parts of a process fit together and can lead to better diagnoses of problems and better decisions. Cross-functional teams lend a diversity of backgrounds to the solving of problems—an important point, as *The Wisdom of Crowds* tells us—and can increase the quality of decisions through divergent thinking and innovation that a specialist may not have the mental framework or training to arrive at individually.

One organization that effectively leverages the collective intelligence of its employees is General Electric's unique jet-engine production facility in Durham, North Carolina. Each engine has approximately 10,000 parts and must be assembled with exacting precision—the lives of millions of air travelers depend on it. GE/Durham consistently produces the highest-quality jet engines in the world. Its people attribute their success to an environment that includes decentralized and collective decisionmaking. GE/Durham only has 1 plant manager for its 170 employees, who work in teams of 15 people and make decisions together.[32]

Hiring at GE/Durham is very selective and takes into consideration an individuals' propensity to help teammates as well as their communication skills, flexibility, coaching ability, and work ethic. Those hired are trained to work in a team environment. At work, they are accountable to the others in the team, and feedback from colleagues is continuous and welcome. Each employee is taught every job by colleagues, and this practice strengthens trust, team awareness, and understanding of how individual tasks fit together. As evidence of this system's success, GE/Durham's people were able to learn how to assemble a new engine in 8 weeks, and then produced it at 12 percent below the cost of a plant that had been producing the same engine for years. Although the plant manager's job is to make sure the plant is making smart decisions as a whole, most decisions are made on the floor by employees or through collaborative teams.[33]

Atlas Container, a Baltimore company, also has a democratic system of decisionmaking that involves employees voting on decisions that directly affect them. Although voting is not always necessary for collective decisionmaking, it is one practice that seems to build morale and trust. Employees have the authority to make changes to processes or systems that they see creative ways to improve and are rewarded for risk-taking. The company is flexible, solves problems, and does not stagnate with systems that work poorly.[34] Atlas boosted its sales from $5.8 million in 1990 to $45 million in 2000 to about $69 million in 2002, with 25 percent per annum growth. Employee retention averages around 85 percent, compared with an industry average of about 50 percent.

Another important practice for collective decisionmaking is the use of teams that cut across functional boundaries. The most innovative and flexible organizations, such as GE/Durham, promote an active cross-pollination of ideas when solving problems. One of the best examples of this concept is a company called Ideo, a highly regarded designer of a wide variety of items from high-technology products to workspaces. Ideo uses interdisciplinary teams of anthropologists, engineers, social scientists, marketers, and designers when developing new products. The teams look at problems from many angles to arrive at sound yet innovative solutions that can give the firm a market advantage.[35]

A key to adaptability is hunger for new information from the market and competitive environment, especially information that challenges conventional wisdom and the status quo. Just as an individual must understand the boundaries of mental models and be receptive to new possibilities to avoid stagnation, so must an organization:

> If an organization seeks to develop . . . life-saving qualities of adaptability, it needs to open itself in many ways. Especially important is the organization's relationship to information, particularly to that which is new and even disturbing. Information must actively be sought from everywhere, from places and sources people never thought to look before. And then it must circulate freely so that many people can interpret it. The intent of this new information is to keep the system off-balance, alert to how it might need to change. An open organization doesn't look for information that makes it feel good, that verifies its past and validates its present. It is deliberately looking for information that might threaten its stability, knock it off balance, and open it to growth.[36]

Semco, for example, not only has an open environment where information sharing is the norm, it shuts down and starts up all over again every 6 months, forcing disruption and requiring everyone to be rigorous and fresh in their activities, such as planning and budgeting. The company takes a fresh look at the organization and questions each business unit's existence. It asks people to justify the existence of their jobs and the top leadership team rotates roles. It avoids the stagnation that bureaucracy can breed by requiring people constantly to be aware of their decisions and how they fit into the organization.

Semco also has been successful in engendering and leveraging collective intelligence and decisions. CEO Ricardo Semler attributes Semco's success to open communication and peer-to-peer relationships, not restricted to need-to-know, parent-to-child approaches of the past. By eliminating unnecessary layers of management and trusting its employees, the company encourages a dynamic system of teams that self-organize, share information openly, and adapt quickly to seize opportunities.[37]

Key Lessons

The increasing use of networking in and out of the military is creating both new opportunities and new challenges, the mastery of which will require more effective use of the human mind. Networked information permits but does not ensure improved reasoning and decisionmaking. At the same time, networking increases the complexity and pace of events by providing torrents of information of uneven quality, faster communication, and fluid interdependencies. The military can learn from wider research and other sectors how to strengthen individual decisionmaking and reform organizations to gain advantages in dynamic, urgent, and complex environments.

Successful organizations understand that information networking alone is no panacea without having people with the right cognitive abilities and a strategy and structure to use them. Such organizations in dynamic markets tend to be choosy and aggressive in hiring people who are not only intelligent but also are open to learning, adaptable, creative, humble yet confident, willing to take responsibility, able to work interdependently, and good at solving problems collaboratively. Hiring individuals with these qualities will produce a workforce capable of self-directed learning, which contributes to adaptability. Decision games can strengthen intuition, even in unfamiliar situations, and enhance understanding of mental models. Likewise, specific training can increase the quality and speed of analytic methods.

In addition to hiring and training, decentralizing authority and entrusting people in the organization to make important and autonomous decisions enable organizations more effectively to harness the cognitive abilities of their people. Furthermore, successful organizations are sharing information and using cross-boundary teams, cross-training, collective decisionmaking, and induced disruption to enhance the quality of problem-solving and performance in operations and markets. Many of these findings can be applied successfully to soldiers, forces, and war.

These lessons in decisionmaking from outside the realm of warfare are worth consideration by the military as the use of networking becomes a commodity among competing entities and the advantage goes to the best thinkers.

Notes

¹ See Gary Klein, *The Power of Intuition: How to Use Your Gut Feelings to Make Better Decisions at Work* (New York: Random House, 2004); and *Sources of Power: How People Make Decisions* (Cambridge, MA: MIT Press, 2000).

² Klein, *The Power of Intuition*, 19.

³ Ibid.

⁴ Bill Breen, "What's Your Intuition?" *Fast Company* 38 (September 2000), 290.

⁵ Charles L. Smith, *Computer-supported Decision Making: Meeting the Decision Demands of Modern Organizations* (Greenwich, CT: Ablex, 1998), 23.

⁶ Klein, *The Power of Intuition*, 7.

⁷ Ibid., 8.

⁸ Ibid., 250.

⁹ Jim Collins, *Good to Great: Why Some Companies Make the Leap—and Others Don't* (New York: Harper Collins, 2001).

¹⁰ Figure derived from information from Keith H. Hammonds, "How Google Grows . . . and Grows . . . and Grows," *Fast Company* 69 (April 2003), 74.

¹¹ Ibid., 74.

¹² David E. Vise, "Google Ends Auction for IPO Shares," *The Washington Post*, August 19, 2004, A1, A9.

¹³ Collins, 42.

¹⁴ Marcus Buckingham and Curt Coffman, *First, Break All the Rules: What the World's Greatest Managers Do Differently* (New York: Simon and Schuster, 1999).

¹⁵ Ibid., 56.

¹⁶ Kevin Freiberg and Jackie Freiberg, *Nuts! Southwest Airlines' Recipe for Business and Personal Success* (Austin, TX: Bard Press, 1996), 66.

¹⁷ Collins, 63.

¹⁸ Doug Krug and Ed Oakley, *Enlightened Leadership: Getting to the Heart of Change* (New York: Fireside, 1994), 59. Oakley and Krug discuss the importance of attitude in hiring and management.

¹⁹ Klein, *The Power of Intuition*, 28.

²⁰ Ibid., 30.

²¹ Methods and tools for supporting the choice of a course of action include cost-benefit analysis, hypothesis testing, systems analysis, and objective weighting.

²² Smith, 35.

²³ Paul Guglielmino and Lucy Guglielmino, "Moving Toward a Distributed Learning Model Based on Self-Managed Learning," *SAM Advanced Management Journal* 66, no. 3 (Summer 2001), 36–43.

²⁴ Freiberg and Freiberg, 113.

²⁵ Ibid., 40.

²⁶ Guglielmino and Guglielmino, 39.

²⁷ Ibid., 39.

²⁸ Ibid., 36.

[29] See Traci L. Fenton, *The Democratic Company: Four Organizations Transforming our Workplace and our World* (Washington, DC, 2002), available at <www.designnine.com/library/docs/other_papers/Democratic_Company.pdf> .

[30] Ricardo Semler, "How We Went Digital Without a Strategy," *Harvard Business Review*, 4 (September 2000).

[31] Brad Wieners, "Ricardo Semler: Set Them Free," *CIO Insight*, April 1, 2004, available at <www.cioinsight.com/article2/0,1397,1569009,00.asp>.

[32] Charles Fishman, "Engines of Democracy," *Fast Company* 28 (October 1999), 174.

[33] Ibid.

[34] John Case, "The Power of Listening," *Inc. Magazine*, March 2003, available at <www.inc.com/magazine/20030301/25206.html>.

[35] Brad Stone, "Reinventing Everyday Life," *Newsweek*, October 27, 2003.

[36] Margaret Wheatley, *Leadership and the New Science* (San Francisco: Berrett-Koehler, 1999), 83.

[37] For a full description of Semco's approach, see CEO Ricardo Semler's writings.

Chapter Eight

Creating a Battle-Wise Force

The Need for a Strategy

Our excursion into the civilian business sector found that many of our main themes resonate across different types of organizations. Sound and timely decisions under pressure demand a combination of intuition and reasoning; delegation of authority and collaborative problem-solving are useful and often necessary in messy, dynamic, information-rich environments; individuals must take responsibility and learn; and training is critical. The basic ingredients of battle-wisdom are as important in other sectors of our society as in the military. Yet the cognitive demands of warfare are especially challenging.

First, soldiers often must make split-second, complex decisions that can have deadly implications and severe penalties for error. Second, because of the ubiquity of videocameras and news teams, the consequences of a military decision may end up on television or the Internet, possibly with international political ramifications. Third, these life-and-death and potentially controversial decisions may be complicated by ambiguities, such as the blurred lines between war and peace, combat and law enforcement, and angry civilians and plain-clothes insurgents. Fourth, soldiers face opponents who are trying to confuse, disrupt, outsmart, and harm them. They must attend to self-preservation and face their own mortality on top of other burdens.

Keeping in mind both the similarities and the differences between warfare and the rest of society, we are ready for a preliminary look at policies the military establishment might pursue to enhance the battle-wisdom of soldiers, teams, and forces. A statement by Army Chief of Staff General Peter Schoomaker about current operations in Iraq sums up quite well the need for battle-wisdom in the modern warfare environment:

> We've had to transition to more unconventional warfare in some highly complex terrain that includes not only cities and towns, but also rivers, valleys, wetlands and desert. . . . I've been most impressed

with the adaptability of our leaders and soldiers, especially the ability of relatively junior leaders to take on roles that were far beyond the traditional scope of a company or battalion commander. Those officers are running towns in Iraq, helping organize and working with civic leaders, making tough decisions day and night, even while conducting combat operations around the clock ... *I think that kind of adaptability and sophistication is something we need to fold back into the batter here as we think about shaping the future Army ...We want an adaptive organization full of problem solvers. We want them to know how to think, not just what to think.*[1] (Emphasis added).

A similar observation comes from an article in *The New Yorker*:

Iraq ... is precisely the kind of unpredictable environment in which a cohort of hidebound and inflexible officers would prove disastrous ... the exigencies of the Iraq war are forcing decisionmaking downward; tank captains tell of being handed authority, mid-battle, for tasks that used to be reserved for colonels ... whatever else the Iraq war is doing to American power and prestige, it is producing the creative and flexible junior officers that the Army's training could not.[2]

There is a crying need for soldiers—especially junior officers and NCOs—who are adaptable, quick to learn, opportunistic, capable of making timely yet sound decisions, and ready for more responsibility. If the U.S. military is to remain operationally superior in networked warfare, it needs people who are more battle-wise, and it needs many such people. The more battle-wise the soldiers are, the better they will be able to function in the chaotic, ambiguous, and perilous situations that await them. And the more battle-wise individuals there are in the military, the more likely it is that U.S. forces will continue to hold a cognitive edge in networked warfare. Because of the importance of local information, initiative, and authority, the U.S. military will not succeed if it relies on a battle-wise few.

Along with improving the cognitive performance of the individual, the Armed Forces should attempt to increase the collective wisdom of military organizations, especially ad hoc combat teams. As noted earlier, the wisdom of crowds can be greater than that of any individual, however smart or capable that person is. In addition, top leadership must exhibit, encourage, and reward the traits that it expects of its employees:

adaptability, anticipation, decisiveness, and a willingness to be accountable for their judgments.

The U.S. military has several policy levers it can use to achieve the goals outlined above. Recruiting strategies can be used to find and attract people with battle-wise abilities and potential. Sorting strategies are important for identifying, grooming, and assigning people capable of effective cognition in operations. Education and training can be used to improve the intuition, reasoning, and specific battle-wise abilities of individual soldiers and teams. Retention policies will remain essential to keeping the requisite numbers of battle-wise people for as long as they are needed. All of these endeavors must reinforce one another; the sections that follow examine how.

Recruiting

One of the most critical factors in creating a top-notch, all-volunteer military is the ability to recruit the right people. Recruiting is the first step in the process of fielding a force of battle-wise soldiers. One of the key challenges facing the military, of course, is that the private sector competes with them for the same pool of talent. This challenge is particularly difficult when it comes to persons who have the cognitive and leadership qualities that allow them to excel in information-rich, complex, pressurized, and time-sensitive environments.

People join the military for a variety of reasons. A survey of the roughly 200,000 people who volunteered for active duty between 1996 and 1998 found that:[3]

■ 30 percent joined to finance their college education
■ 20 percent joined for job training and experience
■ 20 percent joined for the pay and/or travel
■ 10 percent joined out of a sense of duty

Thus, no single policy lever can be used to attract recruits across society. Because networked warfare will require that responsibility be pushed down to lower levels in the military, DOD will need to recruit more battle-wise soldiers in both the enlisted ranks and officer corps. However, the challenges it will face in identifying and recruiting personnel of each type will differ. We begin by analyzing enlisted troops.

Enlisted Personnel

Two criteria are used to identify high-quality enlisted recruits: scores on the Armed Forces Qualification Test (AFQT) and level of education. The former is used to eliminate recruits not likely to succeed in the military: "The AFQT is designed to measure the trainability of potential recruits—more specifically, to identify individuals who are at high risk of not completing the initial training program."[4] It does this by assessing achievement and aptitude rather than pure intelligence (which is why scores on the AFQT tend to rise with age). Current enlistment standards require that 90 percent of recruits have a high school diploma and 60 percent score in the upper half of the AFQT.[5] High-quality enlisted recruits satisfy both criteria.

While education and aptitude as measured by the AFQT do provide some ability to predict military performance, these measures do not capture a number of the intangible attributes of outstanding soldiers.[6] In fact, those intangibles often do not reveal themselves until people are placed in situations that simulate combat: "Research has shown that only on-the-job experience can reveal certain important but previously unobserved aspects of quality, such as effort, reliability, leadership, ability to work as part of a team, and communication skills."[7] To be more precise, research performed by the RAND Corporation has found that roughly 75 percent of a soldier's quality is related to the intangibles that show up during performance of duties, while about 25 percent is related to education level and AFQT score.[8] In addition, the Government Accountability Office has found that it takes 4 years to measure the full performance of a given recruiting class.[9]

Thus, reliable predictions of military performance cannot be made with the current metrics used to recruit enlisted personnel. If this is true in general, it is surely all the more true in gauging the presence of, or potential for, battle-wisdom. Individuals capable of anticipation, initiative, quick reactions, adaptation, and learning in action may be better networked warfighters than others who lack these qualities, even if they have less education and/or lower AFQT scores. Unfortunately, identifying inexperienced people who have such traits is not easy. If it were, the military already would be using those methods to recruit at least some of its soldiers. The reality is that unless and until the Armed Forces are able to develop new tests and methods to evaluate battle-wise traits in potential recruits—a problem that begs for research—it has little choice but to continue using AFQT scores and education as filtering criteria and to seek and develop key abilities later.[10]

Officers

Officers are generally selected in one of three ways: admission to a military academy, Reserve Officer Training Corps (ROTC), or promotion from the enlisted ranks. Each of these paths has its own method for identifying both high performers and those unlikely to survive a career in the military. Although these recruiting methods are different than those used for recruiting enlisted personnel, level of education and achievement are still used as filters, albeit through different means. For example, some service schools rank students based on both academic standing and performance in boot-camp–type professional training.[11] In all three recruiting methods used for officers, the military has some way of observing recruits in either simulated or real operational exercises. As a result, the Armed Forces should be able to do a better job of identifying battle-wise officers than battle-wise enlisted personnel. In fact, research has shown that it may be possible to predict the leadership performance of officers by looking at a small set of cognitive and personality traits.[12]

In general, officers have more years of education than enlisted personnel and thus have more attractive career options and earnings potential in the private sector. Although the military can try to match the starting compensation packages offered by companies, it will be hard-pressed to succeed. One problem is that it has less flexibility in its ability to offer variable forms of compensation, such as bonuses and stock options, because its pay structure is primarily based on rank and tenure rather than skill set, scope of responsibility, and occupation.[13] On the other hand, if DOD simply tried to pay all of its officers at the equivalent private-sector rate, it would end up grossly overpaying some while underpaying others. In any case, this strategy is unaffordable.

Thus, while monetary compensation is important for recruiting officers, by itself it may prove insufficient for attracting the battle-wise many. To recruit high-quality officers today, the Armed Forces focus on the many intangible benefits one can receive from the military. It should continue to do so. In particular, prospective education and training are key levers at the military's disposal.[14] The military can (and does) offer to provide top-notch education and training to officers at little or no cost. This is an attractive proposition for many highly skilled and ambitious people—and, of course, education and training are essential in their own right for developing battle-wise people. Other intangible benefits include the chance to see the world and be of service to one's country. These benefits are explored further in the section on retention.

Lateral Entry

Adjusting recruitment strategies to attract people of generally high cognitive fitness will not guarantee or simplify the way to spot future battle-wise soldiers. Nonmilitary organizations regularly take risks when hiring people with no experience and little basis on which to determine how they will react to stressful, urgent situations. Like the military, companies rely on certain predictors—grades, test scores, measure of aptitude, and fit. And like the military, their results are mixed. Consequently, in addition to recruiting right out of good business and engineering schools, companies in demanding, dynamic markets rely increasingly on hiring experienced people whose past performance can be assessed.

The current U.S. active-duty personnel system is considered to be closed because one must generally enter at the bottom of the hierarchy and climb from there (exceptions to this rule will be discussed). This closed system is different than the civilian personnel system used by DOD, which allows people to enter the defense workforce at middle and senior levels of management.

The current active-duty system is closed for many reasons, but primarily because it is based on an assumption that incoming people are unskilled and inexperienced high school graduates.[15] This approach ignores the fact that the number of high school graduates pursuing college degrees is growing. These potential recruits are gaining education and skills valued in the private sector and expect higher salaries, more responsibility, and more seniority than unskilled high school graduates. With college as a popular option, the traditional pool of high-quality enlisted applicants is shrinking. Simultaneously, the demands being placed on enlisted recruits are increasing: "Requirements are shifting toward enlisted personnel who are knowledgeable decisionmakers who can apply general principles in technical fields, define problems and reach conclusions, and communicate these technical issues effectively to co-workers."[16] A similar problem is found in the officer ranks. For example, due to shortages at the rank of captain, the Army has begun to recruit extra lieutenants and rush them to promotion a year early, whether or not they are qualified—hardly a way to upgrade cognitive capabilities for networked warfare.[17]

The implication of these trends is a potential shortage of battle-wise personnel at the very time when the military needs many more of them. One approach that deserves examination is to expand lateral entry into the military to provide a new stream of future warfighters.[18] Four potential benefits may be gained with such an approach:[19]

- Gaps can be filled in certain fields where young recruits are hard to find or where skill shortages exist due to attrition.
- The potential pool of applicants can be expanded.
- Entrants with backgrounds and known abilities relevant to warfighting can be recruited.
- Training costs can be reduced.

The traditional argument against lateral entry is that recruits are not capable of functioning effectively in the military without starting at the bottom and working their way up through experience and training. While this assumption certainly held true when the vast majority of recruits had relatively low skills and education, today's applicants are both better educated and more highly skilled than in the past. Although some aspects of military training are unique, especially combat training, the private sector now provides many of the general skills that the military teaches its recruits, especially in the areas of combat support and combat service support. Thus, the Armed Forces have the opportunity to bring in experienced people at more senior levels and provide them with military-specific rather than general skills training. This allows the services to get qualified people with experience and proven, relevant abilities into the force quickly. In fact, the Armed Forces do this already with their Reserve forces. The key question is whether this approach should be broadened to people with no prior military experience.

The more the military relies on information networking, and the more that battle-wisdom is required, the more attractive and feasible lateral entry becomes as an option to supplement current recruiting methods. Depending on work experience before joining the military, it may be possible actually to identify battle-wise traits in recruits. For example, if someone were to switch from being a fireman or police officer to being a soldier, the military might be able to assess fairly accurately how that person would perform under urgent, life-and-death circumstances. Of course, recruits who join the military via lateral entry still would be required to undergo basic and advanced military training. While this training possibly could be abbreviated, the real benefit is that the military is likely to end up with more battle-wise soldiers when the training is done than if it had recruited people right out of school.

Both the Army and Navy are currently experimenting with lateral-entry programs for enlisted occupations and having limited success.[20] However, these programs have not focused on attracting soldiers with battle-wise characteristics. On the contrary, they tend to focus on technicians, medical support staff, and musicians. Other programs that focus on lateral entry into the officer corps, such as in the medical and legal corps, appear to be more successful. In addition, the heavy reliance on and excellent performance of reservists in Operation *Iraqi Freedom* and its aftermath demonstrate the value of at least some forms of lateral entry. We believe that the potential benefits of lateral entry justify continued experimentation to determine if it can increase the number of battle-wise soldiers in the military.

Why not recruit college-educated firefighters or police officers with a few years of experience, run them through basic military training, and give them a rank that puts them in charge of recruits who have had no experience dealing with complex, life-and-death situations? Is it not possible that a firefighter or police officer with 5 or more years of experience facing time-urgent, life-and-death decisions is equivalent to a military officer with less than 5 years of experience? Think of the battle-wise potential of a young fire commander. It behooves the military to explore this idea in earnest, especially given the fact that many of its current and future operations will involve counterinsurgency tactics, peacekeeping operations, and urban warfare.

Of course, recruiting people with proven cognitive abilities from high-risk/high-stress professions is not the same as looking for battle-wise potential in the larger population. Perhaps the hardest trait to detect, much less measure, is the courage and cool with which a person will handle receiving or delivering violence. Indeed, the only reliable predictor of future behavior in combat is prior behavior in combat. But this should not rule out lateral entry. Although nothing in the business world compares to warfighting, some of the core cognitive abilities—self-aware intuition-cum-reasoning, fast learning and decisionmaking, anticipation and rapid adaptation—are important in a growing number of civilian sectors, especially where information is plentiful, conditions are ambiguous, problems are constantly changing, and pressures to make timely but sound decisions are great.

Any lateral-entry strategy will have limits, even if it proves successful in the lower ranks. The military has unique training requirements, and the higher one goes in the organization, the more one needs the in-depth knowledge that only can be acquired within that organization or industry. Intensive training can provide a finite amount of information; the rest must be learned through experience seasoned with training. The implications of lateral entry for unit cohesion and morale also need to be examined. Therefore, we suggest a deliberate and experimental approach, addressing important questions before such a system is implemented on a large scale:[21]

- What occupations will be open for lateral entry? Will combat positions be filled?
- What levels of training and experience will be required for lateral entrants?
- What incentives will be provided to attract lateral entrants?
- How will potential recruits be identified?

Lateral entry should be explored as one way to address the looming demand for more battle-wise warfighters. If nothing else, it could be an option when a need arises to increase rapidly the number of battle-wise soldiers—a need that cannot be met quickly enough via traditional recruiting strategies. But even if lateral entry is adopted, it will account for only a small fraction of total recruits. For the most part, the military must still bring in soldiers at the bottom of their hierarchy. Given the challenges associated with trying to identify battle-wise people before they join the military, the Armed Forces have little choice but to focus heavily on evaluating people in their junior years, while still using filter tools like the AFQT to weed out likely drop-outs. This process, called sorting, will be discussed in more detail.

Whichever recruiting strategy is adopted, the military still will need to compete with the private sector for people. It is important that tangible military benefits (base pay, bonuses, retirement pay, and health care) be competitive with, if not necessarily equal to, those found in the private sector. However, we believe that the key to hiring talent away from business and other nonmilitary professions lies with intangible benefits—skills, education, career development, job excitement and satisfaction, and esprit de corps.

Even then, recruiting alone cannot satisfy the need for increasing the number of battle-wise people in the U.S. Armed Forces. Because of the cost of competing for top raw talent and the difficulty of predicting cognitive effectiveness in combat conditions, the military simply cannot follow a Google-like strategy of satisfying its needs by hiring almost exclusively the most talented people with the very best qualities needed. It must augment its recruiting efforts by increasing the battle-wisdom of soldiers already in the fold.

Training and Education

Military education and training serve a number of purposes. Education can provide foundational and contextual knowledge for the military profession, while training can sharpen skills. But both can and increasingly should be used specifically to enhance battle-wisdom for networked warfare to improve the ability of warfighters, both individually and in teams (ad hoc or standing ones) to blend intuition with reasoning to solve complex problems, seize opportunities, exhibit adaptability, and take responsibility for hard choices under extreme pressure and urgency.

The Defense Science Board (DSB) has made similar points about the current ability of the military to prepare soldiers for 21st-century warfare: "transformation of the military will substantially increase the cognitive demands on even the most junior levels of the military. In short, everybody must think. Our current training and educational processes will not adequately prepare our people to cope with these increasing and constantly changing cognitive requirements."[22]

Training

Training in the U.S. military is a success story and an important source of advantage over adversaries. Accordingly, we are able to find numerous examples of training approaches that foster battle-wise abilities and decisionmaking and, therefore, ought to be favored. Current training has both a traditional component and, increasingly, a component that responds to the unfamiliarity and unpredictability of the security environment and operational contingencies. The requirement to train soldiers in standard skills and forms of military operations is not going to disappear. Soldiers must know how to operate equipment, carry out orders, and work together in small and large units.

Beyond this, the requirement to gain time-information superiority in networked warfare implies a growing need for training intuition and reasoning to enhance the abilities we identify with battle-wisdom. As the DSB report stated: "The future will require that more of our people do new and much more complicated cognitive tasks more rapidly and for longer continuous periods than ever before . . . this amounts to a qualitative change in the demands of our people that can not be supported by traditional kinds of training."[23] The DSB is exactly right, but operations in Iraq are showing that the future is now.

As early as 2000, General Eric Shinseki, then-U.S. Army Chief of Staff, opined that about half of a soldier's training was meaningless and "non-essential."[24] Subsequent research by the Army War College revealed that:

> the problem was not "bogus" training exercises but worthwhile training being handled in such a way as to stifle fresh thinking. The Army had so loaded training schedules with doctrinaire requirements and standardized procedures that unit commanders had no time—or need—to think for themselves. The service was encouraging "reactive instead of proactive thought, compliance instead of creativity, and adherence instead of audacity."[25]

U.S. Army Chief of Staff, General Peter Schoomaker, concurs with that assessment but notes, quite rightly, that changes are under way: "In the past, you were measured on how you complied with doctrine and used it to organize and accomplish your objectives. Today, we're designing training scenarios that put people in a continual zone of discomfort . . . that's where we want them. That's how you stretch yourself."[26] The quotation is telling because it implies doubts—healthy ones, in our view—about the value of compliance with doctrine (coming from the chief of the service for which doctrine has always been paramount). Also implicit in General Schoomaker's statement is that change will produce bewildering circumstances for which the trainee must be stretched beyond the familiar. Cognitively intensive and extensive training can help soldiers develop more reliable intuition about warfare through experience in situations with a wide range of patterns, solving various types of problems, and learning to think in strange and confusing circumstances. This will help trainees gain a time-information advantage over the adversary and create room for reasoning to make decisions or verify intuitive choices.

One way to improve decisionmaking is to isolate the *types* of decisions encountered in a certain situation or job function, practice those decisions repeatedly, review the success or failure of those decisions, and make appropriate adjustments. In fact, the U.S. Army National Training Center (NTC) practices a process quite similar to the one just described.[27]

The NTC approach follows several key tenets:

- The best learning comes from the most stressful situations.

■ Learning should be about what matters.

■ The use of hard data eliminates subjective debate about outcomes.

■ Learning requires facilitators who coach rather than lecture.

■ A learning mindset that endures beyond the training exercise should be promoted.

NTC training methods are well-suited for increasing battle-wisdom:[28]

■ Training must be realistic so that soldiers can gain experience recognizing patterns relevant to real-life combat situations.

■ Soldiers are encouraged to experiment, which allows them to gain experience in recognizing what options work in which situations.

■ After-action reports are conducted right after an exercise so that soldiers learn from their mistakes while the experience is still fresh in their mind. Also, through a facilitated approach, the Army forces soldiers to recognize their own successes and failures and determine what other courses of action might have worked. Because they are in charge of their own learning, these soldiers are much more likely to internalize these lessons and incorporate them into their future intuition.

A key feature of such training is to condition people to take chances and allow them to fail. "When you allow people to innovate and lead, you invite failure."[29] This is important for two reasons. First, it is better to have U.S. soldiers learn from failure when training than when facing an enemy on the battlefield. Second, unless taxed to the point of failure, the learning and abilities of trainees may not be fully realized. A further benefit of the NTC approach to training is that it focuses on unit performance as well as individual performance. This is critical in helping the military improve its ability to make collective decisions—one of the capabilities we believe critical for future operations.

A good example of a group whose training methods clearly enhance battle-wisdom is the Navy SEAL.[30] Sea, Air, Land (SEAL) recruits are subjected to multidimensional stress: rather than being faced with a single stress-inducing task (such as a long run), they often must perform two or three difficult tasks simultaneously. For example, recruits are asked to dive into a deep tank carrying three ropes, which they must tie to a rope fixed 6 inches from the bottom. Each knot has to be tied one at a time in a different style. Once they have accomplished this task, recruits must repeat the exercise blindfolded. As a result, recruits learn to handle multiple types of stress while working in an environment they do not control—they learn to "normalize the abnormal." Such exercises are used to build teamwork, leadership, and trust among SEALs. While they involve strength, endurance, and dexterity, these exercises are more about cognitive skills than pure physical skills.

SEALs also train their recruits to failure. This not only weeds out those who lack what it takes to become special operations soldiers, but it also teaches the recruits to recognize their limits and learn how to succeed despite those limits. Although SEAL training is extremely demanding physically, it is worth exploring how the principles and cognitive features used by SEALs and other special operation forces could be incorporated more widely in the less strenuous training methods used for other troops.

Another issue that bears on cognitive development is experimentation. The military has focused a great deal on experimentation in the last decade. It has even created a command focused specifically on joint experimentation.[31] While pursuing experimentation certainly has value, especially to strengthen joint warfare concepts and integration, the current form of experimentation appears to be mismatched with what is required to increase battle-wisdom. This mismatch occurs because experimentation is tied strongly to exercises, and the latter are traditionally used to improve performance in existing operations. Soldiers in exercises do not necessarily gain experience in a wide range of operations that can improve their intuition when faced with novel situations. "In reality, experimental objectives are often at variance with operational requirements—operations each (combatant command) must be prepared to execute . . . because the exercises usually stress traditional operational practices, asymmetric threats, alternative methods of conflict deterrence, and support to peace operations are relegated to secondary importance."[32]

In other words, exercises are often scripted so that specific technologies, doctrines, and tactics can be evaluated in controlled circumstances. Unfortunately, the constraints placed on the opposing forces in such scenarios are often unrealistic (or assume a less-than-clever adversary) and thus undermine the validity of the exercise results. Such a problem occurred with a major exercise called *Millennium Challenge 2002* that was designed to test one of the key concepts of network-centric warfare: effects-based operations. The retired general heading the opposition (Red) force utilized unconventional and innovative tactics that proved so disruptive and destructive to Blue operations that exercise officials reset the game to ignore the effects of Red's unorthodox tactics.[33] The Red commander quit the game, and Blue went on to victory—a victory that failed to determine the limits and weaknesses of Blue problem-solving in the face of a clever, unorthodox enemy.

Our purpose is not to denigrate the utility of exercises and experimentation; both have important roles to play in military training. However, if the military wishes to improve battle-wisdom in junior officers and enlisted personnel, it needs to provide them with a wide range of experiences, including ones that are unorthodox or unusual, in a short period of time. Exercises may be useful in augmenting the training of battle-wise soldiers—by improving their ability to perform in joint operations and international coalitions, for example—but they cannot form the core of cognitive preparation.

One area in particular where experimentation has not been used enough is in evaluating personnel policies themselves. The military should consider creating experimental units to test out new personnel policies. One group that would be a good candidate for such experiments is the Red force that resides at NTC. However, any unit or coherent group of soldiers (such as a career field) could be used to test new approaches for recruiting, training, and retention.[34]

Our analysis to this point has shown that the training methods used by NTC and the SOF community appear well suited to facilitating battle-wisdom. Other training methods, such as large-scale exercises, may not be so suitable. This is not to say that every aspect of training needs to focus on producing battle-wise characteristics, but many soldiers today are not receiving the kind of training they need to be prepared for networked warfare.[35] The military establishment should explore the expanded use of NTC and SOF training methods across the military as a whole.

Another option for strengthening battle-wise thinking and decision-making via training is through the use of simulations. Again, parts of the U.S. military seem to be moving in that direction. One of the most promising systems under development is the Joint Fires and Effects Trainer System (JFETS), which provides extremely realistic three-dimensional simulations of warfare.[36] The motivation behind this effort is to prepare troops for the increasing complexity they are facing in the field:

> The backbone of military training for centuries was rote learning. The goal of the punishing routines and endless drills was to replace thinking with instinct so that at the sound of gunshots, a soldier would automatically return fire. But this kind of schooling, the Pentagon now believes, is inadequate to prepare soldiers for hot

spots like the Sunni Triangle, where it's not enough to be a good marksman. These days, grunts fresh out of basic training must also be versed in the nuances of street-level diplomacy with an increasingly hostile citizenry in densely populated neighborhoods where allies can turn into opposing forces overnight.[37]

The focus of the JFETS training program is to develop soldiers' cognitive skills and decisionmaking ability in high-pressure, time-sensitive environments—in our words, to improve battle-wisdom:

Institute for Creative Technologies programs [such as JFETS] are designed to train the individual soldier in a decentralized, networked model of warfare in which even the lowest-ranking officer can call in an air strike or a tank battalion . . . The Army decided that it needed to think less about educating people on the physics of artillery tubes and start teaching them how to make smart discriminations very quickly in close urban fights—*training in cognitive decisionmaking rather than skills.*[38] [Emphasis added]

Using JFETS, soldiers can train in a wide variety of combat (or noncombat) environments in an extremely short period of time. The simulations are much less expensive than real-life exercises and can be changed daily to reflect up-to-the minute intelligence. This allows the military to improve a soldier's intuition by increasing the number of combat situations experienced in a given time period.

While systems like JFETS are clearly important, use of simulations is not limited to large-scale, multiple-user environments; a wide variety of electronic training aids and devices can be used by individuals as part of either a formal or self-directed learning program.[39] Given that the number of recruits who grew up playing videogames is growing rapidly, soldiers will be increasingly open to utilizing virtual environments as part of their training. In fact, the use of simulations may even have a significant impact on recruiting: "An Army survey of potential recruits found that the game *America's Army* [a videogame in which players take on the role of new soldiers], which cost only $4 million to develop, has made a more positive impression than all the Army's other recruiting initiatives combined."[40]

We have recommended a number of options for improving the training of soldiers for networked warfare. Whatever approaches the Armed Forces take, it is critical to develop a methodology for measuring the efficacy of training methods. Without this information, there is no reliable, timely way to know if a particular training system is having the desired effects. This is especially true, and especially important, for cognition-related training. To this end, the performance of forces undergoing training must be assessed at the individual, unit, and force-wide levels. The latter is important because the individual or small-unit level has too much variation to enable broad conclusions about a military-wide training program to be reached. One way to assess training is to require the Armed Forces to deliver an annual training report card directly to the Secretary of Defense.[41] In addition to providing metrics on service training programs, this approach would help raise the visibility of training and address the chronic underfunding of training: "Training's achievements, its failures and costs are not routinely visible to those with authority over discretionary funding in the Defense Department."[42] If training continues to lag behind other highly visible funding priorities, such as weapons systems, U.S. forces (and the Nation as a whole) will pay the price—especially when faced with well-trained, intelligent, networked adversaries.

Education

The U.S. military has an excellent system of professional military education (PME). Much of the instruction that goes on at the Nation's service schools and joint institutions of learning is built around theory and case studies. This educational approach is quite effective in supporting learning around a solid body of knowledge like doctrine. However, it is less effective for teaching soldiers to develop reasoning skills in uncertain and stressful conditions. In addition, students who exhibit unorthodox thinking often are marginalized or penalized, despite the fact that such thinking may be an explicit goal of the educational system. As a result, students may be susceptible to groupthink. This is partly because soldiers are acutely aware of rank, even in the classroom, and partly because students are acutely focused on grades. Promotions at higher ranks are extremely competitive, and student performance in service and joint schools is a key factor in promotion decisions. Thus, students are often reluctant to jeopardize their academic standing by challenging the system and potentially hurting their grades. As a result, PME sometimes fosters an environment where students are taught to mimic the thought processes of instructors or superiors in the classroom.[43]

Another challenge facing the military, particularly the Army, is that peacekeeping operations call for different skills and knowledge than battlefield operations, but its education system is still primarily focused on warfighting: "While many basic leadership skills carry over from war to peacekeeping, the latter calls for more patience and political and cultural sensitivity. The Army recognizes this challenge, but there is little evidence that it has significantly broadened the education of its rising leaders."[44] Part of the problem is that the decisionmaking styles involved in peace-keeping operations may be quite different from those used in a warfighting situation. Thus, the Army must educate (and train) soldiers to function effectively using two different decisionmaking styles. This is not easy, and once again points to the need for recruiting, training, and retaining those who are best able to switch between these cognitive approaches—that is, those who are battle-wise.

While no easy solution is available to alter the PME system to support different and often opposing needs, some steps appear promising. One option for encouraging innovative thinking is to enforce strictly a policy that all personnel at educational institutions leave rank at the door. This is especially important as a new generation of soldiers who have grown up with computer technology enters the force. Another idea is to change how these schools grade their students, or how the services take academic performance into account in their promotion decisions. Again, the goal is to move away from an overly strong focus on compliance with theory and doctrine to more creative thinking.

A more direct strategy might be to focus the teaching curriculum on the problem of groupthink itself. An example of this approach is found at the Information Resources Management College (IRMC) at the National Defense University. IRMC currently offers a program in Organizaional Transformation that focuses on something called cross-boundary leadership. According to Elizabeth McDaniel, IRMC Dean of Faculty and Academic Programs:

> Cross-boundary leaders must be public servants who are very skill-ful at communicating, coordinating, and collaborating as members of networks across sectors, levels, departments, and agencies. As participants in networks they must foster trust among members, be selfless risk-takers, and effectively focus on intended outcomes to achieve lofty goals as well as concrete goals for their customers . . . and the senior leaders of their own organizations. They must think systematically, influence others without relying on organizational authority, and share responsibility and accountability with other cross-boundary participants. To be effective, cross-boundary leaders

must appreciate cultural and organizational differences, and appreciate, create, and take advantage of networks that rely on technology, management, policy, and people.[45]

It is evident that cross-boundary leaders have the potential to increase the collective wisdom of military organizations. The challenge facing the military is how to integrate cross-boundary leadership principles into its PME curriculum. While the Organizational Transformation program at IRMC can serve as a model, its focus is on government services rather than warfare, and its overall size and scope are limited. More work is needed to adapt this approach to both service and joint colleges. By reinforcing battle-wise training with cross-boundary leadership, the military can strengthen a culture that values the innovative, adaptable, and creative thinking required in today's complex environment.

Retention

If the military proves successful in recruiting, educating, and training battle-wise people, it will still face the challenge of keeping them. To achieve this goal, several obstacles must be overcome. Perhaps the biggest one is that the characteristics that define battle-wise soldiers are in high demand in the private sector. The business world generally offers better pay and more desirable career opportunities than the military, especially for individuals with intuitive abilities. As a result, some of the most capable soldiers are at risk of leaving the military before it has received sufficient payback for the education and training it has provided them, and while it still needs them.

The flip side of this problem is that the existing personnel system makes it difficult for the military to release soldiers it does not need. As a result of these two conflicting challenges, the military has a poor mix of soldiers: "On average between 1999 and 2002, the services had shortages in about 30 percent of their occupations, while they were overstaffed in 40 percent."[46] The options for fixing this problem fall into three general categories: career development, tangible benefits, and intangible benefits.

Career Development

The ability to predict how well a young recruit will perform in military operations based on prior education and test scores is limited. While certain traits can be correlated with high marks in such key military areas as leadership and communication, discerning who has such qualities during the recruiting process is extremely difficult. Often these traits do not appear until a person is placed in a combat situation. Because battle-wisdom consists of elusive qualities, it will be even more difficult to identify green recruits with battle-wisdom potential using existing metrics. Therefore, it is incumbent on the military to identify battle-wise individuals as early as possible in their careers.

This process is called *sorting*. Early sorting to identify battle-wise individuals will allow the military to focus resources not only on developing those individuals but also on retaining them. They could receive higher base pay, performance bonuses, specialized education and training, and career tracks that stress operations. This approach is used for SOF, pilots, and other high-value occupations. The challenge will be to extend it more broadly across the combat forces.[47]

In addition to identifying battle-wise soldiers, the military also should consider highly rigorous sorting around the 10- to 12-year mark, which generally coincides with promotion from O–3 to O–4 in the officer ranks: "[I]t is important that people not reach 10 to 12 years of service without having been selected for their knowledge, skills, initiative, and effort."[48] Because the current retirement system does not vest until 20 years of service, soldiers who make it past the 10- to 12-year threshold tend to remain in the military until year 20. After 12 years, it also becomes harder to justify releasing soldiers based on performance. This is a major cause of overstaffing. Another problem is that after 20 years of service, soldiers, with their cognitive effectiveness potentially at its greatest, often leave the military. After they have vested in their retirement system, many soldiers leave the military to begin a second career in the private sector. Rigorous sorting before soldiers reach the 10- to 12-year mark could target those who will provide value in the future and could help focus on retaining those soldiers beyond 20 years.

According to Bernard Rostker, former Undersecretary of Defense for Personnel and Readiness:

> In the 21st century, the U.S. military needs a competitive up-or-out system in the junior grades with relatively high selection rates, and then stringent selection of only about 30 percent into a career force. Once in the career force, the norm would be very high promotion rates, perhaps 90 percent to O–6 (colonel) rather than the 50 percent of today. Longer tenure and higher remuneration for those selected to join the "career force" should encourage people to stay for a full career

that would end at about 40 years of service. Compensation packages must be structured to motivate the best to stay and encourage those whose potential is limited to leave.[49]

At the same time, if the military creates battle-wise soldiers early, it may not need to retain them for 20 years or more to recoup its investment. If a battle-wise major can make decisions as effectively as a 20th-century colonel, the military may not need as many colonels. As more decisions get pushed down the ranks, the military may become a flatter organization. As a result, retention policy may be used to optimize the mix so that senior positions are no longer overstaffed. However, to do this the military will need to adjust its compensation policies.

Tangible Benefits

One of the primary means that the Armed Forces use to retain people is a mix of tangible benefits that includes pay and nonmonetary benefits, such as housing, medical care, and assistance with child care and education.[50] Because the private sector usually offers higher base salaries and bonuses than the military, one strategy that the military has used to retain its people is to offer generous medical and retirement benefits. Thus, individuals can receive compensation that may rival what is found in business if they are willing to defer those rewards for 20 years.[51] However, the fact that retirement benefits do not accrue until 20 years of service causes some soldiers to stay in the military longer than would be optimal for the force as a whole. This ends up costing a great deal of money and also suboptimizes the allocation of resources, since the funds that could be used to entice "high-aptitude individuals" to remain in the military are tied up with lower-quality soldiers who are staying in the military primarily to obtain their retirement benefit.[52]

The dilemma here is that retirement benefits, triggered at the 20-year point, are one of the advantages the military has over the private sector, which make them an excellent retention tool. The challenge is to retain the best and discourage those who provide little value to the organization. The best way to do this is to sort and target individuals before they reach 10 to 12 years of service. Another problem associated with the 20-year retirement benefit is that it can discourage risk-taking, especially after a soldier has passed the 10- to 12-year mark: "To reach 20 years of service at current rank, the service member must guard against a mistake or misbehavior that would result in demotion or dismissal from service."[53] This is a serious problem if the military desires to grow battle-wise individuals who are willing to take on parochial interests, accept responsibility, and take initiative.

It seems clear that a more flexible retirement system would help the military retain its best and brightest while releasing those people whom it no longer needs. While numerous modifications to the retirement system have been proposed, we believe that the following changes merit serious consideration:[54]

- Adopt a new retirement system that includes full vesting between 5 and 10 years.
- Increase Government contributions (matching funds) for military personnel who participate in the Thrift Savings Plan.
- Remove the one-size-fits-all 20-year annuity in favor of one that allows for differences among occupations.

While overhauling the military retirement system would help in retaining battle-wise individuals, other tangible benefits play an important role in retention. One of the biggest factors that causes prized people to leave the military is the pay differential between it and the private sector. Companies can offer employees higher base pay, significant bonuses (often exceeding base pay at senior levels), and stock options. While it is difficult for the military to match such compensation, it does not have to do so to retain its people. The military offers numerous other tangible and intangible benefits that can compensate for differences in pay; however, monetary compensation in the military must be close enough to what is offered in the private sector so that the differences do not outweigh the other benefits of military service.

The Defense Department has two levers at its disposal in this area: base pay and variable pay (bonuses). One obvious approach to increasing retention (and helping recruitment) is to increase significantly the base pay of all military members. However, the costs associated with a pay raise of this magnitude are so high as to make it nearly impossible. A steeper pay table, by which promotions would produce ever-larger step increases, could offer substantially greater compensation for those who stay and excel.[55] Taking responsibilities into account, military pay tends to become less competitive with the private sector the higher the rank and the longer the service. The present value of future pay is a major factor in retention decisions, especially for highly educated and able individuals who think about long-term financial well-being and know that companies pay senior people well.

The advantage of using base pay as a lever to increase retention is that it minimizes uncertainty and perceived inequity for people in the system: everyone at a given level in a given branch of service will get the same pay. The disadvantage of this approach is that it lacks the flexibility to reward and motivate battle-wise abilities and performance. In contrast, variable pay options such as performance bonuses are extremely flexible, which makes them useful for rewarding strong performance by warfighters facing cognitive challenges. One idea particularly well suited for retention of battle-wise soldiers is capability pay, which is designed to "provide compensation and incentives for superior individual capability, especially current and prospective future leadership potential."[56] Capability pay provides a mechanism for rewarding performance outside the promotion system. As a result, it gives greater flexibility to personnel managers and encourages people who excel in their jobs to remain in the military. Because capability pay can be skewed to become steeper at higher ranks, it can be useful in retaining high value personnel past the 20-year mark.

While variable pay has many advantages, it does carry some challenges. One disadvantage is that variable pay mechanisms may create incentives for me-first behavior. One way to minimize this problem is to tie bonuses to the types of behaviors that improve performance in a networked environment, such as collaboration, sharing, and teamwork.[57] This would create a shared sense of responsibility. On the other hand, assessing team performance can be difficult and will become more so as the composition of combat teams becomes fluid and crosses service lines, as will be the case in networked operations. Another challenge is that systems that reward outstanding behavior must be based on a rigorous set of standards that are perceived as being applied fairly.[58] If this is not done well, the variable pay system may foster a climate of competition and parochialism.

Intangible Benefits

Although efforts to improve retention of battle-wise people should begin with tangible benefits, intangible benefits must not be neglected. The military cannot top private sector pay in competition for effective thinkers and decisionmakers. In fact, focusing primarily on money may prove counterproductive: "the more compensation shifts toward tangible rewards, the more likely it is that professionals who seek the intangible rewards will leave. Their departure, in turn, makes the military a less attractive place for other professionals, creating a vicious cycle away from professionalism toward bureaucracy."[59]

Perhaps the most important benefits the military can offer soldiers are high-quality, low-cost education and training. According to recruits, the top reasons for enlistment in the military are to obtain education and training. By providing its personnel with a rewarding and beneficial experience of lifelong learning, the military can improve recruiting, workforce performance, and retention. People with battle-wise potential will tend to value such a benefit, and the military should give such people priority for educational opportunities. Training and education are especially valued if they are useful outside the military, which is the case for the basic cognitive abilities that make up battle-wisdom. One would think that the better the job the Armed Forces do in developing battle-wise people, the harder it will be to retain those people. However, evidence shows that an organization that gives people marketable skills keeps its employees longer: "There is a strong correlation of psychological commitment and intent to stay (loyalty) with an organization's efforts to make an individual more marketable; the risk of losing employees is greatly increased when organizations fail to provide such opportunities."[60]

Finally, the military can offer people the feeling of camaraderie and satisfaction that comes from working with others to serve one's country. Many soldiers view their military service and lives as a calling and their relationship to the Nation as a compact. Of course, the Nation must do its part to ensure that its soldiers are not taken for granted or put in harm's way unnecessarily. The stresses and strains of frequent, long, and dangerous deployments can take an immense psychological, emotional, and financial toll on both active-duty and reserve personnel and their families.[61]

In sum, when it comes to attracting, developing, and keeping battle-wise people, no magic answers—at least not affordable ones—are available to the military. Still, several ideas seem promising and deserve consideration: lateral entry; training for problem-solving in unfamiliar, ambiguous, and urgent circumstances; educational emphasis on analytic skills and cross-boundary collaboration; rigorous sorting before the 10- to 12-year mark; a more flexible retirement system; a steeper pay scale; and increased use of variable monetary incentives, such as bonuses. Further research is needed to determine whether and how these ideas should be pursued.

Reforming Command and Control

If the decisionmaking abilities of each networked warfighter can be expanded by the sorts of measures suggested, it stands to reason that the performance of U.S. military forces in networked warfare would be enhanced by involving lots of them, as individuals and in teams, in decisionmaking and increasing the demands on each. This requires action to discourage top-down decisionmaking.

In the traditional perspective, centralization of authority reflects less a willful hoarding of authority than a natural distribution of responsibility from the strategic to the operational to the tactical planes. However, that three-plane model has been disturbed by the growing speed, fluidity, and ambiguity of warfare, which blur and compress these planes, increase the significance of tactical decisions, and reward horizontal, peer-to-peer collaboration. The operational effects of tactical choices and the strategic effects of operational choices flow freely and quickly across these porous boundaries.

In addition, the traditional view of decisionmaking presumed that the force commander and staff (at the top or in the back, depending on one's perspective) would possess more relevant information and experience than the warfighter (on the bottom or at the edge). This view is no longer valid for many situations. The beliefs of the old culture—that experience counts above all and that top commanders are better informed than lower ones—are being battered by the geopolitical and networking revolutions. In unfamiliar conditions, and with data easily shared, neither experience nor information at the top/back necessarily trumps cognition at the bottom/edge. Mobilizing the battle-wise many is made necessary by change, ambiguity, and complexity, and made practical by networking.

Decentralization must, of course, accompany any attempt to have more battle-wise warfighters throughout the force, up and down the ranks. Entrusting junior officers and NCOs in the field to make quick, critical, and sound judgments demands that they can intuit reliably and reason efficiently, are aware of their analytical and experiential strengths and limitations, can learn in action, and are adept at adaptive decisionmaking. Therefore, decentralization of decision authority to take advantage of networking will both require and reward efforts by the Armed Forces to build up these cognitive strengths.

Extending decisionmaking to more warfighters on the network depends on devising command and control architectures that permit the shifting of authority downward and outward. But reform is not just about decentralization because networking not only informs warfighters but also makes them interdependent by expanding options for collaboration. Command and control architectures should accommodate the need for units and decisionmakers throughout the networked force to support and be supported by others, regardless of geography, service boundaries, and normal operational command boundaries. Permitting local and peer-to-peer problem-solving weakens vertical control and demands strong and open horizontal and diagonal links, which do not easily fit with rank and structure. However, even as commanders delegate their traditional authority, they are indispensable in managing interdependencies, settling disputes, and allocating scarce resources.

A formidable obstacle to any reform, particularly one involving command and control, is that the effort to organize for joint operations is now frozen just beneath the level of the joint force commander. So-called component commanders are little more than service commanders with responsibility to coordinate the operations of the forces of their service with those of other services. Thus, while information networking is permitting deeper integration and horizontal collaboration, joint command and control only exists at the upper echelons. Over the long term, this will not do for achieving deep, operational integration, which is the essence of networking in any field. In fact, recent operations already reveal weakness in joint command and control on the battlefield.[62]

In the transition from control by the few to empowerment of the many, it may be useful to have a few enduring rules to govern who should make what decisions:

■ First, commanders should communicate an understood envelope within which subordinates may and should operate, defined by mission objectives, limits, and available resources. The limits of authority should be predicated on whether decisions (including bad ones) taken by subordinates may have consequences (including unintended ones) outside their envelopes.

■ A second rule could be that the decisionmaking authority of an individual is contingent on that individual's having at least as much information as a superior commander does. Even in a networked environment, it will sometimes be the case that headquarters has some information bearing on a tactical situation that cannot be rapidly shared with the warfighter—for security reasons, for example.

■ Third, any individual who does not feel equipped with the intuitive and reasoning powers to make a sound decision should unhesitatingly seek and receive advice, guidance, and, if need be, orders. Self-awareness of limitations is a strength, not a weakness.

■ Fourth, if an individual wishes to use forces that are not under his or her regular command or have not been placed at his or her disposal, the commander of those forces must concur in whether and how they are to be used.

To illustrate, we can revisit the major leading the ambushed force in the African peace-enforcement operation. If he decided wrongly and failed in his mission or suffered numerous casualties, yet his decision did not have significant consequences outside his envelope of responsibilities, then giving him decision authority was probably the correct move. If, however, his misjudgment exposed other units to a larger losing battle or jeopardizes the peace of the province as a whole, perhaps the decision should not have been his to make. Similarly, if the major's superior knows but cannot communicate, for whatever reason, that the window to carry out the unit's mission is closing, the superior may have to tell the major that pulling back is not an option. If the major or his superiors are convinced that the situation is more complex and dangerous than he is prepared for, it might be best not to risk failure even within his envelope of responsibility. Finally, if the major wishes to call in reinforcements from another unit, and if they have not already been placed at his disposal, then he will have to seek a decision from the officer commanding those other forces or, failing that, seek intervention from higher command.

Intangible qualities—self-awareness, trust, and an educated feel (not some rigid structure) for who is best placed to decide what—are of great and growing importance in networked warfare. Other than combat itself, both training exercises are potentially the most effective way to inculcate forces with these qualities. Whereas Navy culture traditionally has stressed delegation of authority, autonomy, and accountability, the Army has stressed reliance on fellow soldiers—obviously reflecting the difference between a ship at sea and a company in the field. A fusion of the former's trust in "the skipper" and the latter's trust in "buddies" is needed for battle-wise forces in networked warfare.

Developing Battle-Wise Teams

Developing collective intelligence for military operations will not be easy. Networking certainly will help by providing good communication, shared awareness, additional information, and decentralization of authority. However, because the forces deployed for an operation and present on an operational network are fluid and dependent on circumstances, it is not yet clear how to choose the assortment of units to be exercised. Obviously, more joint exercises are warranted. However, it will take considerable resources and organizational flexibility to plan exercises involving various combinations of, for example, SOF, bombers, unmanned airborne sensors, land forces, missile-carrying submarines, and aircraft carriers.

We hesitate to advocate wholesale application of the wisdom-of-crowds principle to military problem-solving. After all, the warfighters of a force are not all faced with the same tactical problem but, rather, myriad ones. Yet this approach could have merit in the case of a group of people who organize to face a common problem, which is the very idea of ad hoc, cross-boundary, military-operational teams. Assuming they are accommodated by flexible command and control, such teams can bring to bear diverse perspectives on common problems—precisely the conditions in which collective wisdom excels. Thus, crowd-wisdom could translate into battle-wisdom under certain conditions.

Networking theory, more or less confirmed by practice, suggests that ad hoc teams will self-organize to deal with common problems, enabling an organization continuously to optimize its resources despite uncertainty and change. Take the case of the major and the ambushed unit. All else being equal, forming a team with other networked unit commanders for the sake of deciding whether to get his unit out of harm's way or engage in a firefight would offer little gain in the quality of the decision and significant risk to its timeliness. Yet, once ground and gunship support arrive, it may make more sense for the several officers concerned to discuss and even decide together whether to eliminate the ambushers or instead brush them aside and get on with the mission.

If ad hoc teams can be crowd-wise, the question remains of how to make them battle-wise. A reasonable starting point could be what appears important for the individual warfighter: a provisional decisionmaking approach to gain time and information; self-awareness of collective experiential and analytical limits; the ability to learn in action; and an emphasis on the abilities that create operational time-information advantages—anticipation, rapid decisionmaking, opportunism, and rapid adaptation.

This is daunting enough for individuals; its achievement by teams, however promising in theory, will be very hard. All in all, the concept of collective wisdom in military operations—creating it as well as using it—requires much more thought, research, and experimentation. But it should not be dismissed simply because it seems to defy the principle of unity of command. The test will be whether it is possible to involve multiple decisionmakers without running the risk that no decisionmaker emerges at all.

Notes

[1] "The Buzz," *Government Executive*, December 2004, 18.

[2] Dan Baum, "Battle Lessons: What the Generals Don't Know," *The New Yorker*, January 17, 2005, 42–48.

[3] Bill Owens, *Lifting the Fog of War* (New York: Farrar, Strauss & Giroux, 2000), 49–50.

[4] M. Rebecca Kilburn, Lawrence M. Hanser, and Jacob Alex Klerman, *Estimating AFQT Scores for National Education Longitudinal Study (NELS) Respondents* (Santa Monica, CA: RAND, 1998).

[5] Paul F. Hogan, "Overview of the Current Personnel and Compensation System," in *Filling the Ranks: Transforming the U.S. Military Personnel System*, ed. Cindy Williams (Cambridge, MA: MIT Press, 2004), 30.

[6] James R. Hosek, "The Soldier of the 21st Century," in *New Challenges, New Tools for Defense Decisionmaking*, ed. Stuart E. Johnson, Martin C. Libicki, and Gregory F. Treverton (Santa Monica, CA: RAND, 2003), 191.

[7] Hosek, "The Soldier of the 21st Century," 196.

[8] James R. Hosek, interview by author, Washington, DC, October 8, 2004. *Quality* is understood as the level of job match between the member and the military. It is often correlated with promotion speed; the faster someone gets promoted, the higher their "quality." For more information, see James R. Hosek and Michael G. Mattock, *Learning About Quality: How the Quality of Military Personnel is Revealed Over Time* (Santa Monica, CA: RAND, 2003), xi.

[9] Richard J. Koucheravy, *Whence the Soldier of the Future? Recruiting and Training for the Objective Force* (Fort Leavenworth, KS: United States Army Command and General Staff College, School of Advanced Military Studies, 2001), 24.

[10] For further discussion of the challenges associated with identify high-quality recruits (both enlisted and officers), see Koucheravy, 23–24.

[11] An interesting insight into the life of cadets at West Point is provided in David Lipsky, *Absolutely American: Four Years at West Point* (Boston: Houghton-Mifflin, 2003).

[12] Paul T. Bartone, Scott A. Snook, and Trueman R. Tremble, Jr., "Cognitive and Personality Predictors of Leader Performance in West Point Cadets," *Military Psychology* 14, no. 4 (2002), 321–338. This study focused on the leadership performance of cadets during their 4-year tenure at West Point. If the predictors in this study are shown to be robust when applied to officers after they graduate from West Point, we recommend that they be tried out on enlisted personnel.

[13] Thomas M. Strawn, "The War for Talent in the Private Sector," in *Filling the Ranks*, 75. There are some notable exceptions to this statement. Officers in certain highly competitive and in-demand occupations such as aviation, medicine, and nuclear engineering receive significant sign-on and retention bonuses. We discuss this point later in the chapter.

[14] One of the most successful tools for recruiting college-bound individuals is loan repayment. For more information on this and other policy choices related to recruiting from the college market, see Beth Asch, Can Du, and Matthias Schonlau, *Policy Options for Military Recruiting in the College Market: Results from a National Survey* (Santa Monica, CA: RAND, 2004).

[15] Donald J. Cymrot and Michael L. Hansen, "Overhauling Enlisted Careers and Compensation," in *Filling the Ranks*, 120.

[16] Ibid., 121.

[17] Cindy Williams, "Introduction," in *Filling the Ranks*, 2.

[18] *Lateral entry* can be defined as allowing recruits to enter the military at a rank other than E–1 or O–1 (the bottom of the pyramid).

[19] Levy et al., *Expanding Enlisted Lateral Entry: Options and Feasibility* (Santa Monica, CA: RAND, 2004), xiii.

[20] For more details, see Levy et al.

[21] Ibid.

22 Defense Science Board Task Force on Training for Future Conflicts, "Memorandum for the Chairman, Defense Science Board" (Washington, DC: Office of the Undersecretary of Defense for Acquisition, Technology, and Logistics, July 9, 2003).

23 Defense Science Board Task Force on Training for Future Conflicts, "Final Report" (Washington, DC: Office of the Undersecretary of Defense for Acquisition, Technology and Logistics, 2003).

24 Baum, 42.

25 Ibid.

26 *"The Buzz,"* 18.

27 See Richard Pascale, "Fight. Learn. L*E*A*D," *Fast Company*, August/September 1996, 65.

28 See Gary Klein, *Sources of Power: How People Make Decisions* (Cambridge, MA: MIT Press, 1999).

29 Baum, 43.

30 The information on SEAL training was provided by Rear Admiral Raymond C. Smith, USN (Ret.), in phone interview by author, Washington, DC, January 10, 2005. Although our discussion here focuses only on Navy SOF, the other services use similar methods to train their SOF units.

31 More details available at <www.jfcom.mil>.

32 Thomas M. Cooke, "Reassessing Joint Experimentation—Out of Joint," *Joint Force Quarterly*, no. 28 (Spring-Summer2001), 102–105.

33 See Sean D. Naylor, "Fixed War Games? General says Millennium Challenge 02 'was almost entirely scripted,'" *Army Times*, August 16, 2002; and Malcolm Gladwell, *Blink: The Power of Thinking Without Thinking* (New York: Little, Brown and Company, 2005), 145. For an in-depth discussion of both this exercise and the opposing force commander's military philosophy, see Gladwell, 99–46.

34 For more details on this idea, see Stephen Peter Rosen, "Implementing Changes in U.S. Military Personnel Policy," in *Filling the Ranks*, 295.

35 See Defense Science Board Task Force, "Final Report."

36 See Steve Silberman, "The War Room," *Wired*, September 2004, 151–155, 171–173. This system was the brainchild of the Institute for Creative Technologies (ICT) at the University of Southern California. ICT is a collaboration between DOD, film and gaming companies, and Silicon Valley.

37 Ibid., 153.

38 Ibid.

39 Koucheravy, 38. For a good discussion of the use of tactical decision exercises and simulations to improve both leadership training and decisionmaking, see Major J.B. Vowell, *Between Discipline and Intuition: The Military Decision Making Process in the Army's Future Force* (Fort Leavenworth, KS: United States Army Command and General Staff College, School of Advanced Military Studies, 2004), 53–55.

40 Shawn Zeller, "Training Games," *Government Executive*, January 2005, 46. According to Koucheravy, the DOD budget for recruiting advertising was $265 million in fiscal year 2000. Thus, the cost effectiveness of America's Army is striking.

41 Defense Science Board Task Force, "Final Report," 70.

42 Ibid., 7.

43 There are obvious exceptions to these points. We are making rather broad generalizations based on our observations and experiences because we believe they contain a grain of truth that needs to be expressed.

44 Thomas L. McNaugher, "Refining Army Transformation," in *The U.S. Army and the New National Security Strategy*, ed. Lynn E. Davis and Jeremy Shapiro (Santa Monica, CA: RAND, 2003), 299.

45 Elizabeth A. McDaniel, "Facilitating Cross-Boundary Leadership in Emerging E-Government Leaders," *Electronic Government*, Vol. 1, forthcoming.

46 Williams, "Introduction," in *Filling the Ranks*, 2.

[47] Details on sorting strategies can be found in Hosek, "The Soldier of the 21st Century," 204–207.

[48] Ibid., 204. If they reach the 10- to 12-year threshold, officers tend to remain in the military until they hit 20 years (barring serious issues with performance). Enlisted personnel have a less obvious threshold because their promotions are less tied to time-in-grade.

[49] Bernard Rostker, "Changing the Officer Personnel System," in *Filling the Ranks,* 160–161. Rostker also served as Undersecretary of the Army and Assistant Secretary of the Navy for Manpower and Reserve Affairs.

[50] Elizabeth A. Stanley-Mitchell, "The Military Profession and Intangible Rewards for Service," in *Filling the Ranks,* 94.

[51] Of course, this depends on a number of factors, including rank, career field, and years of service. It is also worth noting that many soldiers who retire at 20 years enter the private sector and thus end up with the best of both worlds: generous retirement and health benefits from the military plus private-sector salaries and bonuses. This is another reason why military personnel who make it past the 10- to 12-year mark tend to stay until 20 years and then leave; they know they can begin second careers at a relatively young age while still taking advantage of military retirement and health benefits.

[52] Hosek, "The Soldier of the 21st Century," 204.

[53] Ibid.

[54] See Strawn, 88.

[55] This idea is examined in James Hosek and Beth Asch, *Air Force Compensation: Considering Some Options for Change* (Santa Monica, CA: RAND, 2002).

[56] Ibid., xv.

[57] See Strawn, 78.

[58] Hosek and Asch, xvi.

[59] Stanley-Mitchell, 94.

[60] Strawn, 89.

[61] See James R. Hosek and Mark Totten, *Does Perstempo Hurt Reenlistment? The Effect of Long or Hostile Perstempo on Reenlistment* (Santa Monica, CA: RAND, 1998).

[62] Richard Kugler, Michael Baranick, and Hans Binnendijk, *Anaconda's Lessons for Joint Operations* (Washington, DC: Center for Technology and National Security Policy, forthcoming).

Chapter Nine

Recommendations and Conclusions

This book began with the point that U.S. military forces, by harnessing information technology and adopting networking principles, can improve speed, flexibility, precision, integration, and overall performance in a wide range of contingencies, provided they overcome institutional inertia and service parochialism. Yet using the same principles and exploiting the diffusion of IT, real and potential U.S. adversaries as diverse as al Qaeda and China can counter these U.S. advantages, according to their own strategies. As a consequence, even though the United States and its democratic partners have taken a dominant lead in military-networking, they cannot assume it will last. As enemy forces become more aware, lethal, dispersed, and integrated, they will become less visible and less vulnerable to U.S. forces; at the same time, U.S. forces, though networked and superior, will become more visible and more vulnerable. With their elusive networks, cellular structure, and fluid tactics, terrorists and insurgents in Iraq today provide a preview of this danger. The skill and speed with which transnational terrorists, despite their technological poverty, already are using networking—linking cells, conducting operations, and disseminating messages of death—point to a network-versus-network future.

In that future, the operational edge will lie with the side that uses brainpower to make better sense and use of information—anticipating enemy moves, making quick decisions, seizing opportunities, learning rapidly, and adapting in action. It will not be enough for U.S. forces to have superior firepower, sensors, and bandwidth. If they are to hold a clear operational advantage over any adversary in any contingency, they must be able to create and exploit time-information superiority. By doing so, despite being more vulnerable to the networked sensors and precision weapons of the enemy, U.S. forces can increase the exposure of enemy forces and reduce their own. Gaining and holding such an edge requires battle-wisdom.

More precisely, the abilities of U.S. military personnel, up and down the ranks, to solve complex and unfamiliar operational problems in the face of danger, urgency, ambiguity, and information overload must be improved. Given that human rationality is often wanting when wrestling with reality, especially in such critical conditions, this is a formidable challenge. No single measure or simple formula will answer that challenge. It will take a multifaceted strategy spanning personnel policy, command and control principles and processes, and intelligent, ad hoc combat teams. It also will take openness to learn from nonmilitary spheres and organizations that are pioneering the exploitation of networking to improve performance, including better decisionmaking under pressure.

As we have noted, people in the military and other high-stakes, high-intensity professions tend to rely mainly on their intuition when time is short. However, today's fluid security environment and unfamiliar operational circumstances make experience less relevant and, thus, intuition less reliable. At the same time, reasoning, though obviously preferred when ample time and information are available, typically is underutilized in battlefield decisionmaking precisely because it is time-consuming. With the tempo of military campaigns increasing, mainly because of networking, intuition inevitably remains essential. But the messier the world and its battlefields become, the more important it is to improve reasoning, using the information processed and shared by networks. Simply put, the aim is to make intuition more reliable and reasoning more time-efficient, thus enhancing military operational problem-solving and lending an increasingly important advantage. But how?

The U.S. military can be proud of the quality, intelligence, and attitude of its people since the birth of the all-volunteer force after the Vietnam War. Nonetheless, the way the U.S. Armed Forces recruit, organize, and prepare people for the mental demands of operations can be improved. This is not a criticism of U.S. military personnel systems but a recognition that the demands on those systems are shifting because of the mix of networking and messiness. Today's personnel systems, with the exception of some innovative training approaches, have not been designed to maximize the particular bundle of decisionmaking abilities that are rising in operational importance: anticipation, decision speed, opportunism, rapid learning, and adaptation. Nor is professional military education especially geared to decisionmaking methods that are both time- and information-sensitive and that integrate intuition and reasoning. Now, for strategic reasons, these qualities and methods should be stressed. With practical steps, not drastic change, they can be.

In this spirit, we conclude this book by recommending concrete actions to be taken, including additional research to be pursued, if U.S. warfighters, forces, and decisionmaking methods are to become more battle-wise. If the U.S. military is to exploit its superior networking to preserve operational and strategic advantages, it will have to consider a variety of policies and measures aimed at developing warfighters who are more battle-wise; command and control systems that empower and support more battle-wise warfighters; battle-wise, self-forming teams; and fast-adaptive decisionmaking methods that integrate intuition and reasoning. In places, we applaud existing efforts being undertaken by DOD and suggest more of the same. In other areas, we recommend new approaches to be examined, tried out, or implemented. In both cases, our recommendations are only a starting point. For one thing, it is likely that a number of efforts are under way that we are not aware of. For another, the very wisdom of crowds that we describe herein may find solutions we have not discovered. With these possibilities in mind, the following recommendations are provided:

1. Explore the use of lateral entry to increase the number of potential battle-wise soldiers entering the military.

2. Leverage training techniques used by the National Training Center and Special Operations Forces, as well as virtual training environments, to strengthen battle-wisdom across the Armed Forces.

3. Orient professional military education more toward cross-boundary leadership and analytic discipline, and remove disincentives to taking risks and presenting radical ideas.

4. Strengthen retention by performing more rigorous sorting before the 10- to 12-year mark in military careers, and develop new retirement and pay policies that will help the services keep high performers while releasing others.

5. Stress distributed and horizontal decisionmaking in command and control, and experiment with joint command and control arrangements that permit deep integration and spontaneous teaming.

6. Foster cross-boundary team cognition and problem-solving.

7. Launch research and analysis efforts to understand better the challenges and opportunities involved in developing a battle-wise force that can create and exploit time-information superiority.

1. Explore the use of lateral entry to increase the number of potential battle-wise soldiers entering the military.

Both the public and private sectors agree that one of the keys to successful organizational performance—be it thriving in financial markets, providing services to citizens, or fighting wars—is finding and hiring the right people in the first place. This is easy to say but hard to do, especially in the public sector. For one thing, the military cannot entice valedictorians with stock options or promises of vast riches and corner offices. For another, it is extremely difficult for DOD to know in advance how people will think and perform when faced with urgent life-and-death decisions in the midst of combat.[1] The best indicator by far of how people will handle intense cognitive pressure in the future is how they have handled it in the past.[2] The problem for the military is that it recruits people primarily straight from high school or college. Scant information is available to help DOD determine which recruits can become battle-wise warfighters.

If companies make bad hiring decisions, long-term harm is not irreparable; if the military makes bad decisions, lives may be lost and national security may be affected. One option DOD should consider for improving its ability to recruit battle-wise people is to see if it can develop and use some sort of cognitive profile and/or screening method to indicate key battle-wise qualities: anticipation, opportunism, reasoning under pressure, willingness to take responsibility and make decisions, ability to learn in action, flexibility, and self-awareness. It is not apparent to us whether such an approach is feasible, or how it would be used. Further research is clearly necessary.

A more promising approach for increasing the number of battle-wise soldiers entering the military is to use selective lateral entry. A number of civilian professions currently place individuals in situations that bear a resemblance to aspects of networked warfare: urgency, life-and-death decisions, complexity, unfamiliarity, and ambiguity. It is possible that recruiting such individuals, giving them essential military training, testing them, and then placing them in positions of leadership will prove to be a successful way of increasing the battle-wisdom of the junior ranks. Although a number of issues clearly would need to be explored and addressed, we believe this option should be given a thorough test before being rejected as either too hard or too disruptive, as skeptics might claim.

Most people agree that recruitment of enough people with high battle-wise aptitude could be achieved by offering substantially higher pay. While pay is a powerful recruiting tool, it also is extremely expensive.

Trying to increase cognitive abilities substantially throughout the military by increasing pay across the board would be cost prohibitive. However, monetary compensation can be tailored to focus recruiting (and retention) on targeted specialties and high-value recruits. Such strategies include steeper pay tables, loan-repayment programs, and enlistment bonuses.

Evidence has shown that the military can attract valued recruits—at least in the case of much-sought-after IT specialists—by offering education and training that will pay off whether the person stays in or leaves the military.[3] In fact, the chance to get broadly relevant, top-notch education and training for free is one of the major reasons people join the military. Thus, providing such education and training has multiple benefits: it aids in recruiting people with battle-wisdom, it increases the battle-wisdom of people who have been recruited, and it helps retain battle-wise soldiers who might otherwise leave just as their value increases sharply, between 5 and 10 years of service. Clearly, education and training should figure centrally in a total strategy to build a battle-wise force.

2. Leverage training techniques used by the NTC and SOF, as well as virtual training environments, to strengthen battle-wisdom across the Armed Forces.

The training methods of the past—those focused more on doctrine and standard operating procedures—will not serve soldiers well in networked warfare. Instead, training programs should focus on developing cognitive abilities by placing soldiers in situations where they are forced to think quickly, adapt, seize opportunities, and learn on the fly. In the words of Army Chief of Staff General Schoomaker: "In the past, you were measured on how you complied with doctrine and used it to organize and accomplish your objectives. Today, we're designing training scenarios that put people in a continual zone of discomfort . . . that's where we want them. That's how you stretch yourself."[4]

We have identified the training programs used by the NTC and SOF as exemplars for the Armed Forces generally. This is not to say that these two programs should be adopted wholesale without change. Each service has unique requirements, and different specialties within each service may need specialized training above and beyond what we recommend. However, the fundamental principles followed by the NTC and SOF programs—realistic situations, freedom to experiment, training to failure, analysis after the fact to learn what worked and what did not, and multidimensional stressors—apply to all forces that take part in networked warfare.

Another recommendation concerns the use of simulations. To develop battle-wisdom in junior officers and enlisted personnel, these soldiers must be exposed to a wide range of experiences in a short period of time. Exercises are certainly a necessary component of this training program, but they are not sufficient for two important reasons: they are too expensive and difficult to develop and use with the frequency required, and they often do not provide the freedom of action necessary to develop battle-wisdom. One solution to this problem is to make greater use of simulations. Simulations can present the kind of messy, ambiguous, unfamiliar, and complex situations faced by soldiers in networked warfare at a fraction of the cost of and with greater flexibility than full-scale exercises. They can be modified quickly (in a matter of days or even hours) to represent new environments for the soldiers. The Armed Forces already are using simulations with much success and placing more emphasis on this aspect of training. This will help build battle-wisdom, especially to the extent that it emphasizes those particular abilities that can yield time-information superiority.

3. Orient PME more toward cross-boundary leadership and analytic discipline, and remove disincentives to taking risks and presenting radical ideas.

The U.S. military has a fine and enviable system of PME. However, much of the instruction that goes on at service schools and joint institutions of learning is designed to support learning of a body of knowledge rather than forcing soldiers to think analytically and unconventionally. In fact, students who exhibit unorthodox thinking are often penalized because their ideas may run counter to the body of knowledge being taught. Two other factors that inhibit risk-taking in educational institutions are that soldiers are acutely aware of rank in the classroom and extremely focused on grades. Promotions at higher ranks are extremely competitive, and student performance in service and joint schools can be a factor in promotion decisions. The net result of those factors is that PME sometimes fosters an environment in which students are taught to mimic the thought processes either of their instructors or of their superiors in the classroom.

We recommend a number of steps that may help remedy the situation described above. First, military educational institutions need to enforce strictly a policy that "rank does not enter the classroom" (which includes the professor). Second, schools need to reconsider how they grade their students. If battle-wisdom matters, then PME must foster and support the development of those key attributes, and it cannot do so if it discourages risk-taking and creativity. A final and related challenge is that the services must reconsider the role of academic performance in their promotion decisions. If students need to maintain "A" averages to get promoted, and if getting an "A" means demonstrating doctrinal and by-the-book thinking, then PME may actually hinder rather than help the development of battle-wise personnel.[5]

The Nation's military colleges and universities also need to stress cross-boundary thinking and foster cross-boundary collaboration. Doing so will actively develop in students the types of skills and thought processes needed to excel in the collaborative networked warfare. Finally, placing greater weight on how to think and less on what to know would pay dividends. Learning or refreshing analytic methods and adaptive decision principles as part of PME would be beneficial, especially in the course of confronting students with the requirement to solve complex and unfamiliar problems.

4. Strengthen retention by performing more rigorous sorting before the 10- to 12-year mark in military careers, and develop new retirement and pay policies that will help the services keep high performers while releasing others.

Given the difficulty of identifying battle-wisdom potential among green recruits, it is important to sort new soldiers as early as is meaningful. This will allow the services to identify individuals with high levels of battle-wisdom and take steps both to leverage their skills and place them on specific career tracks. This will also not only improve the performance of the soldiers in the field, but will help with retention by allowing the military to tailor its policies, such as pay, training, and career development. A similar approach is used for SOF today; we recommend applying it to a broader set of skills that includes those associated with battle-wisdom.

It is especially important to conduct rigorous sorting before people reach the 10- to 12-year mark because those who make it past that point tend to stay for a full 20 years. The reason for this is that military retirement benefits do not vest until someone has put in 20 years of service. The consequence of this policy—and the challenge it presents to recruiting battle-wise soldiers in junior ranks—is that the military has too few soldiers of the sort it needs in the junior ranks and too many in the senior ranks. We have addressed the former problem earlier. The latter problem can be addressed through the use of both rigorous sorting and the creation of a more flexible retirement system.

For example, if soldiers were able to fully vest in their retirement system somewhere between 5 and 10 years of service, then fewer soldiers would feel the need to "hang on" until 20 years (which costs DOD a great deal of money and does not necessarily yield a positive return on investment). By the same token, the military needs to develop incentives for keeping people it does value beyond 20 years (at which point many soldiers leave the military to have second careers in the private sector). It can accomplish this goal via a mix of both tangible benefits—such as base pay, variable pay, and retirement contributions—and intangible benefits, such as education and training. In particular, a steeper pay scale (by which compensation increases significantly at higher ranks and/or years of service) could play a major role in improving the attractiveness of remaining in the military. Not only is a steeper pay scale likely to result in increased retention, it also may help in recruiting. Thoughtful young people base career decisions at least partially on long-term earning potential; one of the major lures of joining professional services companies such as law firms, accounting firms, and consulting firms is the prospect of making partner, at which point compensation increases dramatically. Steeper military pay tables, properly implemented, could play an important role in creating a battle-wise force.

5. Stress distributed and horizontal decisionmaking in command and control, and experiment with joint command and control arrangements that permit deep integration and spontaneous teaming.

Depicting the warfighter instead of headquarters in the center of a network has value. After all, the ultimate purpose of most military networks, like most other networks, is to satisfy the needs of the user for access to information and opportunities to collaborate. Users include lieutenants, lieutenant generals, and everyone in between. Such a perspective would foster the design and testing of network command and control according to the responsiveness of the warfighters' information needs, cognitive performance, and, ultimately, operational effects. It could help expose what technical investments would most improve user accessibility to systems that host critical data, many of which were not designed to support networked operations. It also could reveal how best to promote the sort of distributed, horizontal, and interdependent decisionmaking that can exploit information and cope with fast-changing circumstances.

It is especially important to conduct tests and exercises to learn how responsive various command and control schemes are to warfighter collaboration and problem-solving needs, as well as to experiment with ways to improve responsiveness. Only exercises, besides actual operations, will permit realistic experimentation with and refinement of distributed and horizontal, collaborative, interdependent decisionmaking.

In addition to improving command and control, exercises and experimentation will aid in the measurement of how well efforts to build battle-wisdom are doing. Distribution of information and authority can produce observably better operational results—but only if those receiving the increased authority have the cognitive abilities to use it well. Put differently, if those at the warfighting edge (or network center) are not up to solving increasingly difficult battlefield problems, despite having more and better information, it may be better to leave control with the force and component commanders. However, through measured exercising and experimentation, work to improve warfighter cognition and distributed decisionmaking can be managed as an integrated undertaking.

The U.S. military must figure out how to achieve deep integration—pushing joint command and control downward—such as by having deployable joint command cells available to support tactical commanders in the prosecution of fluid cross-service operations. Joint headquarters could designate which tactical commander would have joint command for a given mission or task well below the component commander level. As this is done, the battle-wisdom of junior officers must and can be improved so that they can be entrusted to organize and lead joint operations at their level.

Progress along these lines is critical to the goal of taking full advantage of the cognitive potential on any given network. As we have noted, local commanders should and can have an information advantage—over both their adversaries and their own superiors—by virtue of being able to pull whatever information they need from the network while also having unique, sensory-based, immediate knowledge. Unless they have the authority to make decisions, however, this will be for naught. Moreover, unless decisionmaking authority for integrated operations is pushed out to the warfighter, tactical decisions will be made by remote generals and staffs—the opposite of the objective of engaging more brainpower closer to the action, as well as getting better strategy from the generals.[6]

Of course, the further out and deeper authority is pushed in the context of joint operations, the harder it may be to know or resolve who is in charge and who is expected to support whom at any moment and for any task. This presents all the more reason to intensify exercising and experimenting with teaming and horizontal decisionmaking—the concomitant of decentralization that has enabled much of the corporate world to improve productivity.

The smart-pull principle of information distribution must be relentlessly pursued.[7] In recent years, civilian IT has made great strides toward meeting this challenge. Before long, the command "search" should be tantamount to summoning whatever useful stored information exists anywhere that is meant to be accessible. Importing the latest methods into the military for both new networks and existing ones must be a priority. The growing commitment of DOD to network-centric enterprise services technology is an important step toward better accessibility, even in a heterogeneous network environment. Once again, smart pull is vastly more complicated for the warfighter than for the Internet user. While the latter's information needs can be reasonably anticipated and met, the former's information needs often will be shaped by ambiguous and dynamic circumstances that defy anticipation and even complete communication—for example: the scale, distance, closing speed, nature, and intent of an enemy force; the weather; tactical objectives; fatigue or other impairments; and operational tempo. Despite the immensity of the challenge, however, this is the right direction to take in military command and control and networking.

6. Foster cross-boundary team cognition and problem-solving.

Although our analysis emphasizes the capabilities of individual soldiers, soldiers rarely fight alone. It is equally important for groups of fighters, independent of service, to be able to form into battle-wise teams or units. Although the concept of self-forming teams that exercise collective intelligence to accomplish the task at hand may seem like a recipe for anarchy, the notion of *self-synchronization* lies at the very heart of network-centric warfare. Thus, as DOD refines operating concepts for networked warfare, the services must concentrate on developing the protocols and habits of team-forming and decisionmaking.

How can DOD best develop battle-wise teams? Once again, the key lies in training and experimentation. We already have discussed some of the difficulties and limitations associated with creating truly realistic, large-scale joint exercises that allow Blue forces to face clever, networked Red ones. However, such exercises, done properly, are an excellent tool for improving cross-service, and even cross-national, collective decisionmaking. That said, in military operational contexts, collective intelligence is better aimed at solving specific microproblems rather than large-scale challenges. Cross-structural teams should be organized around the logic of the operational challenge at hand, be it an ambush, the sudden discovery of a terrorist cell, better-than-expected enemy air-defense capabilities, or the like.

In the context of larger service or joint force exercises, greater effort should be placed on fostering self-formation of teams to cope with particular problems. If problems must be solved by the command hierarchy taking tactical control, the integrative potential of networking and collective thinking will be lost. Once a team forms, the challenge is to reach rational conclusions exploiting all available information and, as important, the diverse perspectives of the members. The advantage of building and using team-based collective intelligence is that it augments the warfighters' networked knowledge and immediate sensory awareness with the problem-solving potency of diverse perspectives.

7. Launch research and analysis efforts to understand better the challenges and opportunities involved in developing a battle-wise force that can create and exploit time-information superiority.

We make these recommendations fully mindful that they will not be embraced by busy policymakers and careful bureaucracies without further study. When it comes to such complex issues as recruiting, retention, PME, training, and distributed and horizontal decisionmaking, shifts in direction have to be carefully and critically examined before being blessed and implemented. For example, tailoring recruitment and retention goals and tools depends on calculations of complicated econometrics and shifting elasticities governing how people respond to monetary and other incentives. We hope that the proposed recommendations will trigger such research and analysis, and we realize—and agree—that policy change must await such work.

Some questions especially deserving of further study follow:

- What are the prerequisites adversaries must meet to be able to exploit networking militarily, and how and when might they meet them?
- As adversaries are able to exploit networking, what will be the effects on U.S. military operational performance?

- What are the strategic and security implications of these operational effects?
- Is it possible to develop a profile for potential recruits that identifies those who have battle-wise abilities?
- How should recruiting standards and strategies be altered to target people with these aptitudes and abilities in sufficient numbers?
- Should quantitative and qualitative retention goals change with the advent of networking, the decentralization of authority, the flattening of organizations, and the stress on people with key cognitive abilities? What mix of personnel policies can help the military meet those goals?
- How can the PME system place more emphasis on developing and recognizing battle-wise abilities and the decisionmaking methods that utilize them to best effect?
- How should training and exercising be sharpened to make intuition more reliable and reasoning more time-efficient in operational problem-solving?
- How should command and control networks, structures, and procedures be designed and developed to improve the distribution of authority and the efficacy of horizontal decisionmaking?
- Can and should the concept of the wisdom of crowds be applied to military decisionmaking? If so, how is it best developed, evaluated, and inculcated into the force?
- How can the goal of and progress toward improved cognitive performance in networked operations be measured?
- How well is the current U.S. military culture aligned with that goal, and how can the alignment be improved?
- How can the effectiveness of new training and education programs and personnel policies be measured?

Conclusions

Because the goal described in this book—exceptional minds making sound decisions in the heat of networked warfare—has strategic significance, we will conclude at that level.

Every so often, the plane of military competition shifts. By the beginning of the 20th century, grand fleets and continental armies had become less important as industrialized military power moved to the fore. Germany, Great Britain, the United States, and Japan stood apart and competed fiercely—at times, violently—based on their ability to combine industrial productivity and military excellence. In the aftermath of World War II, nuclear and aerospace power eclipsed mechanical power. Only two superpowers could assemble the massive resources and expertise needed to compete in these realms. Toward the end of the Cold War, IT entered the military domain. One of the superpowers—the one that lacked consumer and capital markets—could not compete technologically or keep its empire intact geopolitically. With each shift, the field of competitors shrank as fewer and fewer could marshal the requisite economic and technological resources for military purposes.

At the beginning of the 21st century, the networking of information and forces offers a potent combination of awareness, precision, speed, dispersion, and integration in military operations. With its excellence in IT, engineering, and advanced military systems, the United States is and will remain the leader. Unless and until an authentic peer challenger emerges, head-to-head warfare and tit-for-tat competition with the United States are not wise. Yet paradoxically, the scope for military competition has been reopened by this development. Information-network technologies tend to be inclusive, not exclusive. With widely available information services, readily accessible global network infrastructure, abundant bandwidth, and rapidly spreading technical know-how, growing numbers of states and nonstate groups will be able to use information networking to improve their operational awareness, precision, speed, dispersion, and integration.

Although enemy forces will not be able to rival U.S. military network capabilities, this does not mean they cannot be shrewder and quicker than U.S. forces in exploiting information. Military history shows that new eras of competition are not prejudged by the outcomes of preceding eras. In a world of networked warfare and networked opponents, the risks associated with U.S. military action could increase, the certainty of decisive success could decline, and the willingness of the United States to be the principal provider of international security could be thrown into doubt.

Operationally, the loss of networking monopoly will translate into increased visibility and vulnerability of U.S. forces. However invulnerable and invincible those forces may have seemed in the invasions of Afghanistan and Iraq, their subsequent struggles in Iraq may preview things to come. As the transformation of U.S. forces proceeds, an aim of paramount importance must be the attainment of time-information superiority. Networking is necessary but not sufficient to attain such superiority, especially as opposing forces start linking sensors and shooters with precision weapons and dispersing their forces. Convergent vulnerability requires thinking and decisionmaking that compounds the complexity faced by the adversary and compresses the time the adversary has to understand, reason, and react. Surprise, suddenness, and reaction speed become as important as firepower and bandwidth to lethality and survivability.

Whether the concern is with al Qaeda in the near term, China in the long term, or some other wily and determined adversary along the way, it is imperative that the United States sharpen the cognitive abilities and decisionmaking methods of its military personnel—battle-wise individuals and teams alike. The U.S. military needs to increase the number of minds in its ranks that have the ability to make decisions under the pressures of war in the increasing complexities of a networked environment. With its exceptional people, proven personnel systems, and excellent military education institutions, the United States has all the basic ingredients it needs to develop superlative battle-wise forces.

Networked adversaries also will surely grasp the pivotal importance of time-information and the cognitive skills needed to enhance it at the expense of U.S. forces and interests. Al Qaeda is already behaving as if it understands this well, even if in terms that might not resonate with us. Thus, just as cognition could become the key factor in networked warfare, so will it become the new plane of strategic competition, whether with global terrorists or rising powers.

Do America and Americans have inherent advantages in such competition—other than wealth, which does not guarantee better cognition and decisionmaking? RAND researchers contemplating this question opined: "In such a contest, volunteer military personnel drawn from an open, educated society like that of the United States would appear to have the advantage over a stove-piped military embedded in an authoritarian state, but the blinding pace of social, cultural, and technological change in China strongly suggests that this conclusion will not always remain true."[8] We may think of the Chinese, with their tangle of Mandarin and Maoist roots, as not being able to produce and motivate risk-taking, responsibility-taking, battle-wise people the way Americans can. Yet China eventually could become a sufficiently open and educated society that competition with the United States on the cognitive plane would not be lopsided at all.

Al Qaeda and other terrorist groups with strategic ambitions may pose greater dangers than China. The philosophies of such people are, of course, the opposite of those we associate with human enlightenment, rationality, and responsibility. But this does not mean they are incapable of good intuition, effective reasoning, anticipation, opportunism, reaction speed, and learning-in-action. All indications are that al Qaeda has and values people with precisely these abilities, which apparently can coexist with fanatical religious beliefs. Indeed, fanaticism undoubtedly helps in recruitment. In sum, battle-wisdom may be no more confined to the democratic "enlightened" West than IT has proved to be.

Ultimately, a democratic society should have an edge in the capability to find large numbers of volunteers with the education, creativity, objectivity, and willingness to learn and lead essential to battle-wisdom—but only if it sees the need and makes the effort. This book has presented a case that gaining and holding cognitive superiority is of strategic importance—indeed, the stuff of grand strategy—in the coming age of networked warfare. By the same token, the effort to achieve that superiority should be strategic—with the senior attention, coherence, and call on resources implied by the term.

In time—and it may be soon, if the spread of Internet use is any indication—many states and nonstate groups will come to embrace not only networking principles, such as smart pull and distributed decision-making, but also the emphasis on improved cognition that flows from those principles. Will this represent some ultimate plane of warfare and strategic competition—the end of an inexorable escalation from fists to clubs to spears to arrows to guns to tanks to missiles to information networking to minds? Other than going back to invest ever more on the lower planes, with questionable returns on investment, the future of conflict and defense, and hence of war and peace, is likely to be based increasingly on the human brain.

As networking creates the opportunity to gain operational and strategic advantage with the human mind, American security interests and responsibilities argue for taking that opportunity. Yet this raises some philosophical questions: Does improving the ability of the brain to wage war constitute human progress? Do we want our smartest people to be warfighters? Is it good or bad for global security that warfare becomes more intelligent? This study has side-stepped such questions. But it seems reasonable that the better the judgment of the soldiers of the forces that stand for peace and security in the world—by which we mean those of the United States and its democratic allies—the less likely aggression will succeed, the more judicious military decisions will be, and the less destructive, indiscriminate, and, perhaps, frequent warfare will be. On balance, as the quality of thinking takes on greater importance in strategy and warfare, the societies that favor and foster reason, objectivity, and individual genius should have an advantage—if only they seize it.

Notes

[1] Malcolm Gladwell, "Personality Plus," *The New Yorker*, September 20, 2004, 42–48.

[2] James R. Hosek, "The Soldier of the 21st Century," in *New Challenges, New Tools for Defense Decisionmaking*, ed. Stuart E. Johnson, Martin C. Libicki, and Gregory F. Treverton (Santa Monica, CA: RAND, 2003), 196.

[3] James R. Hosek et al., *Attracting the Best: How the Military Competes for Information Technology Personnel* (Santa Monica, CA: RAND, 2003).

[4] James Kitfield, "Army Chief Struggles to Transform Service during War," *Government Executive*, October 29, 2004, 18.

[5] This is not to say that doctrine and theory are not important—they are. The issue is whether they take precedence over "out of the box" thinking that is required to deal with today's complex and ever-changing environment.

[6] One potential consequence of pushing decisionmaking down the ranks is that the military's overall structure may flatten a bit; fewer people will be required at the top while more people will be required at the bottom or edge. As a result, service retention strategies may begin to place even greater emphasis on quality over quantity at the higher ranks, for example O–5 and above.

[7] David S. Alberts and Richard E. Hayes, *Power to the Edge: Command and Control in the Information Age* (Vienna, VA: CCRP, 2003)

[8] James C. Milevenon et al., *Chinese Response to Military Transformation and Implications for the Department of Defense* (Santa Monica, CA: RAND, 2006), xvii.

Afterword

Linton Wells II

The information revolution is transforming our societies and our way of life. DOD leverages these critical developments to build its own networks, empower its people, and improve the effectiveness of its operations and business practices. However, in the course of a recent senior-level defense review, an analyst asked an important question: "Are network-centric operations evolutionary or transformational?"

In one sense, network-enabled capabilities are evolutionary in that they draw heavily on commercial technology developments. But technology alone is not enough. To realize the revolutionary potential of the network, several factors need to evolve together. These include doctrine, organization, training, material, leadership, personnel, and facilities. When these changes are synchronized, truly dramatic, transformational leaps are possible.

Much has been written about the material and technological aspects of network-centric operations and warfare: how much information can be delivered, how much data can be stored, how much bandwidth is available. But it is not by accident that most of the critical changes mentioned above deal with people—how they are trained and organized, how they are recruited and led, what kinds of doctrine they follow. This is precisely the focus of *Battle-Wise*. As the authors point out, in future network-versus-network engagements, "the operational edge will lie with the side that uses brainpower to make better sense and use of the information."

In developing this thesis, they focus on the cognitive domain, which is precisely where path-breaking research is most needed. A few years ago I had the chance to see a Marine Corps Amphibious Assault Vehicle that had been modified into an advanced command and control platform. The vehicle's 4 radios had grown to 12, including some satellite systems, and the troop compartment was filled with computer displays covered with glowing icons. As I looked at this, I thought, "If I had this vehicle, I'd command differently than I would if I had a conventional one with just four radios and no computers." Shortly after, I met a friend who was a Marine general and explained what I had seen. He replied, "Lin, this terrifies me, since it gives senior officers the power to destroy the greatest advantage of our armed forces, which is the initiative of our junior officers and senior enlisted personnel. Now commanders can meddle all the way down to the lowest tactical level."

The response surprised me, since I had thought what I had seen was a significant step forward. But, on reflection, I realized that the Navy had gone through a similar experience years earlier, when it introduced a wide area data link that could share tactical dispositions across entire theaters. It was not long before fleet commanders began calling battlegroup commanders thousands of miles at sea to suggest adjustments in tactical dispositions. Not only was this resented afloat, but it also was not the best way to fight. Eventually reason prevailed, and commanders began using the same kind of "control by negation" that had been adopted for tactical situations in the "composite warfare commander" concept.[1] The point is that through a mix of doctrine, discipline, and training the Navy was able to strike the right balance between operational needs and technological possibilities. This will be possible in other warfare domains also, such as land, air, space, and cyber.

One of the strengths of *Battle-Wise* is its focus on the interaction between leadership and technology. This, too, often is neglected in writings on modern systems, but it is absolutely essential to understanding the network-centric world.

The book makes the point that "the best indicator by far of how people will handle intense cognitive pressure is how they have handled it in the past." The importance of realistic training in combat environments has been shown over and over again, from boot camp to *Top Gun*. Now we must teach our commanders to "Fight the Net" like a weapon system, and not just treat it as an administrative adjunct to modernized business practices or operational procedures. At the same time, we must move beyond static concepts of protecting our networks by perimeter defenses to "mission assurance"—being able to accept damage and fight through it to complete the mission, irrespective of the kind of attack the network is under. This is not primarily a technological problem. Most of the vulnerabilities in network defense stem from mistakes made by people. It is the responsibility of the commander to see that his or her command is battle-wise in both the physical and cognitive dimensions of network warfare.

The principle of network-centric operations is straightforward: the availability of the network lets information be distributed so that participants can establish and share situational awareness. In concert with command intent, this shared awareness allows units to self-synchronize their actions rather than wait for orders from each echelon of a hierarchical command structure. We are moving away from the traditional "push" of information by those who have "owned" it in the past and are giving users the ability to "pull" the information they need from the network.

These changes will require innovative thinking at every level of an organization. People who have "owned" information must be willing to share it, even as network designers must provide enough security to let this sharing happen responsibly. Moreover, the network itself must be built to be "battle-wise" in the sense that it must meet the needs of the full range of operational users from senior commanders on stable, high capacity networks to junior personnel in bandwidth-constrained, dynamic tactical situations. Many kinds of information need to be brought together, especially operational and intelligence data.[2] Forces in contact must know enough to define the operational pictures they need, to instruct their "discovery services" how to pull the information they need from the net, and to act on it effectively. Senior commanders must be willing to accept that they will have less direct control over subordinates' activities. One may even ask if "command and control" should be redefined in the network-centric environment.

A key advantage conveyed by network-enabled capabilities is agility. The existence of a network in 2002 allowed Special Forces units on horseback in Afghanistan, with radios and GPS receivers, to call in precision-guided munitions carried by B–52s for close air support. Virtually no system in those engagements was used in the way it had been designed. The future is likely to be similarly uncertain. It is worth remembering that many of today's Army units were designed for armored warfare against the Soviet Union in Europe, the Air Force was shaped for air superiority over the inter-German border, and much of the Navy was built to keep open North Atlantic sea lines. Yet they have adapted, and will continue to adapt, helped by the flexibility brought by the net and by the battle-wise operators and designers who can improvise on short notice as changing circumstances dictate.

The importance of adaptability is reinforced by the length of the DOD planning and acquisition cycle. The Department plans, programs, and budgets over a 6-year time horizon (the Future Years Defense Program, or FYDP). The Defense Program Projection extends another 10 years beyond the FYDP. It is worth noting that these 16 years are longer than from the time of the Wright Brothers' first flight to the end of World

War I. In a period where commercial product cycles are measured in months, and the global political environment also changes rapidly, individual defense programs typically take more than a decade to develop and acquire, and may be in the inventory for decades more. Both our warfighting and our business practices, and the people who implement them, need to be nimble enough both to keep pace with the competition and to take advantage of emergent opportunities. Networks encourage such adaptability, which is one reason why "net-ready" is the only key performance parameter mandated for any new DOD capability. But network-centric concepts cannot simply be repeats of those used to defeat the Taliban or the Iraqi army. Battle-wise thinking will have to address the challenges of asymmetric conflict in urban canyons as much as conventional war in the open desert or precision strikes against Tora Bora.

Taking full advantage of these networked capabilities may require quite different skills from those we see today in the U.S. military. Not only should we encourage the use of new tools like blogs, wikis, and collaborative spaces to flatten hierarchies and speed decisionmaking, but massive multiplayer online games also are converging with future command and control systems. Our young people are well suited to operate in these environments. But, as this book notes, adversaries and potential adversaries also are enlisting people with network-centric skills, and the global pool of technologically literate individuals from which they can draw is growing.

The authors have described several innovative ways to attract, train, and retain the kind of information-age workforce that we will need to prosper in this new era. This kind of long-term thinking needs to be folded into the DOD planning process, even as we also adapt our operational approaches and business practices to move from the industrial age to the information age. Congress recently authorized the National Security Personnel System, which offers exceptional flexibility for managing the civilian workforce, although DOD managers are only beginning to understand its implications. Many battle-wise ideas may be adaptable to this environment.

In sum, *Battle-Wise* is a timely and valuable book. It addresses serious problems that our security forces will face and provides reasonable and often provocative recommendations on how to solve them.

Notes

[1] When battle speeds became too fast for a single flag officer to control an engagement, command was decentralized to an "anti-air warfare commander," "anti-submarine warfare commander," "strike warfare commander," etc. The governing principle was "control by negation," wherein the

overall commander stood back and let the engagement be conducted by his subordinates unless he chose to weigh in.

[2] In fact, DOD is in a unique position to provide leadership in the area of information fusion because it deals not just with intelligence stovepipes, operational data, and business flows, but with many different kinds of information. But this also requires specially trained people to articulate and execute the mission.

Glossary of Key Concepts

Battle-wisdom is the effective melding of reliable intuition and efficient reasoning to gain time-information superiority in complex, intense, and possibly confusing networked warfare. Battle-wisdom demands self-awareness, the abilities to anticipate, decide quickly, seize opportunities, and adapt in action, and the willingness to lead and learn. In practice, it also depends on implementation of the smart-pull principle of information management and delegation of authority. Battle-wisdom may be employed to increase the exposure time of enemy forces and reduce that of one's own forces—a key factor in tipping the balance of vulnerability to one's advantage.

Cognition is the mental process of knowing, including aspects such as awareness, perception, reasoning, and judgment.

Collective intelligence is a phenomenon in which the errors of individuals tend to cancel out one another as numbers increase, leaving the average to be that much better. Collective intelligence only works if there is ample diversity and independence of views among the participants, so that the full range of experiences, perspectives, and information of the many are in play, resulting in a better answer than if the solution is based on the experience, perspectives, and information of only a few, even if they are of above-average intelligence.

Convergent vulnerability is a condition in which the ability of each of two opposing forces to operate safely is offset by the ability of the other to find and destroy it. The stronger, better networked force would, of course, be less vulnerable than and better able to find and destroy the inferior one. But that superior force would be more vulnerable and less effective than if it were opposed by a non-networked force, all else being equal.

Intuition is the power to attain direct knowledge or cognition without evident rational thought and inference. Intuition enables unconscious problem-solving, which may seem simple—like having a hunch or "gut feel"—but in fact involves complex brain activity. Although intuition is more than learning from repetitive experience, it does appear to function more effectively in dealing with familiar rather than strange circumstances. Research shows that decisions in combat, as in other intense and urgent circumstances, are made mainly using intuition by drawing on experience and going with familiar solutions, rather than analyzing and comparing the costs and benefits of multiple options.

Reasoning is the power of comprehending, inferring, or thinking, especially in orderly, rational ways.

Self-awareness is the knowledge of the origins, assumptions, biases, and limits of one's mental models and experiences.

Time-information is the product of time and information. In decision-making, time can be made more valuable if it is used to gather, evaluate, and exploit information. In turn, the ready availability of credible and useful information can permit better use of time, compensate for a lack of it, and, in effect, make it last longer. The quality of a decision improves as a function of both time and information. The enhancement of time-information, thanks to networked information, should improve the quality of urgent decisions.

About the Authors

David C. Gompert co-authored this book while a Distinguished Research Professor in the Center for Technology and National Security Policy at the National Defense University (NDU). Previously, he was the Senior Advisor for National Security and Defense at the Coalition Provisional Authority, Iraq. Mr. Gompert has held senior policy and executive positions at the State Department, the National Security Council, the RAND Corporation, and in the information technology industry. He has published extensively on international affairs, national security policy, and technology. His books include *Right Makes Might: Freedom and Power in the Information Age* and *Mind the Gap: A Transatlantic Revolution in Military Affairs.* Mr. Gompert holds a Master of Public Affairs degree from the Woodrow Wilson School, Princeton University, and a Bachelor of Science degree in engineering from the U.S. Naval Academy.

Irving Lachow is a Senior Research Professor in the Information Resources Management College at the National Defense University. Dr. Lachow has extensive experience in both information technology and national security. He has worked for Booz Allen Hamilton, the RAND Corporation, and the Office of Deputy Under Secretary of Defense (Advanced Systems and Concepts). Dr. Lachow received his PhD in Engineering and Public Policy from Carnegie Mellon University. He earned a Bachelor of Arts in Political Science and a Bachelor of Science in Physics from Stanford University.

Justin Perkins is a research associate in the Center for Technology and National Security Policy. Previously, he served as chief operating officer for World Blu, Inc., a consulting firm pioneering the field of organizational democracy, and as co-founder and director of Afrique Profonde, a human rights organization in Congo. He also has been involved with several small businesses and served for several years as a water resources administrator for the State of Colorado. Mr. Perkins holds a Masters of Business Administration from the University of Colorado and a Bachelor of Arts in history and world perspectives from Principia College.

Rear Admiral Raymond C. Smith, USN (Ret.), is a former Deputy Commander for the U.S. Special Operations Command and former Commander of the Naval Special Warfare Command.

Linton Wells II is Principal Deputy Assistant Secretary of Defense (Networks and Information Integration) and the Acting Deputy Assistant Secretary of Defense for Spectrum and Support.